Brassica Breeding and Biotechnology

Edited by A. K. M. Aminul Islam,
Mohammad Anwar Hossain
and A. K. M. Mominul Islam

Published in London, United Kingdom

IntechOpen

Supporting open minds since 2005

Brassica Breeding and Biotechnology
http://dx.doi.org/10.5772/intechopen.87490
Edited by A. K. M. Aminul Islam, Mohammad Anwar Hossain and A. K. M. Mominul Islam

Contributors
Mohammad Akmal, Arif Hasan Khan Robin, Subroto Das Jyoti, Naima Sultana, Lutful Hassan, S.A. Dwivedi, Bahaderjeet Singh, Ryo Fujimoto, Ayasha Akter, Mst Arjina Akter, Hasan Mehraj, Takeru Itabashi, Tomoe Shindo, Naoki Chiba, Masaaki Osaka, Naomi Miyaji, Junji Miyazaki, Jogendra Singh, Ali Bandehagh, Zahra Dehghanian, Mohammad Anwar Hossain, Robert Henry, María Elena Cartea, Antonio De Haro-Bailón, Sara Obregon-Cano, Fernando Cámara-Martos, Francisco Rubén Badenes-Pérez, Kailash Chandra, Subhash Chand, Om Prakash Patidar, Rajat Chaudhary, Ranjit Saroj, Vijay Kamal Meena, Omkar M. Limbalkar, Manoj Kumar Patel, Priya P. Pardeshi, Prashant Vasisth, Parbodh Chander Sharma, Vijayata Singh, Lelika Nameirakpam, Ajay Tomer

First published in London, United Kingdom, 2021 by IntechOpen
IntechOpen is the global imprint of INTECHOPEN LIMITED, registered in England and Wales, registration number: 11086078, 5 Princes Gate Court, London, SW7 2QJ, United Kingdom
Printed in Croatia

British Library Cataloguing-in-Publication Data
A catalogue record for this book is available from the British Library

Additional hard and PDF copies can be obtained from orders@intechopen.com

Brassica Breeding and Biotechnology
Edited by A. K. M. Aminul Islam, Mohammad Anwar Hossain and A. K. M. Mominul Islam
p. cm.
Print ISBN 978-1-83968-696-2
Online ISBN 978-1-83968-697-9
eBook (PDF) ISBN 978-1-83968-698-6

For EU product safety concerns:
IN TECH d.o.o., Prolaz Marije Krucifikse Kozulić 3, 51000 Rijeka, Croatia, info@intechopen.com or visit our website at intechopen.com.

Meet the editors

Prof. Dr. A. K. M. Aminul Islam is Professor of Genetics and Plant Breeding at Bangabandhu Sheikh Mujibur Rahman Agricultural University (BSMRAU), Gazipur, Bangladesh, where he is also a director of research. He received bachelor's and master's degrees from Bangladesh Agricultural University (BAU), Mymensingh, Bangladesh, and a Ph.D. in Chemical and Process Engineering from Universiti Kebangsaan Malaysia. Dr. Islam is the author of 120 articles published in nationally and internationally reputed journals, twenty book chapters, and four books. He is an editorial board member and referee for several national and international journals. He is also the general secretary of the Plant Breeding and Genetics Society of Bangladesh, the seminar and research secretary of JICA Alumni Association of Bangladesh, and a lifetime member of several professional societies. Dr. Islam developed and released nineteen varieties of different crops for commercial cultivation by farmers. He supervised twenty-two MS and two Ph.D. students as major professor and forty MS and two Ph.D. students as a committee member. His major areas of research are the development of hybrid vegetables, canola-grade *Brassica napus* using a CMS system, and salinity-tolerant rice, as well as renewable energy research with *Jatropha curcas.*

Dr. Mohammad Anwar Hossain is a professor in the Department of Genetics and Plant Breeding, Bangladesh Agricultural University (BAU), Mymensingh, Bangladesh. He received his BSc in Agriculture and MS in Genetics and Plant Breeding from BAU. He also received an MSc in Agriculture from Kagawa University, Japan, in 2008, and a Ph.D. in Abiotic Stress Physiology and Molecular Biology from Ehime University, Japan, in 2011 through a Monbukagakusho scholarship. As a JSPS postdoctoral researcher in 2015–2017, he worked on isolating low-phosphorus, stress-tolerant genes from rice at the University of Tokyo, Japan. His current research program focuses on understanding physiological, biochemical, and molecular mechanisms underlying abiotic stress tolerance in crop plants and the generation of stress-tolerant and nutrient-efficient plants through breeding and biotechnology. He has more than sixty peer-reviewed publications to his credit and has edited twelve books, including this one.

Dr. A. K. M. Mominul Islam is a professor in the Department of Agronomy, Bangladesh Agricultural University (BAU), Mymensingh, Bangladesh. He received both bachelor's and master's degrees from BAU. Dr. Islam obtained a second master's degree in Physical Land Resources from Ghent University, Belgium, and a Ph.D. in Plant Allelopathy from United Graduate School of Agricultural Science, Ehime University, Japan, with his dissertation "Allelopathy of five Lamiaceae medicinal plant species." Dr. Islam also completed his post-doctoral research at the Department of Horticulture and Landscape Architecture, Purdue University, Indiana, USA. Dr. Islam is the author of seventy-four articles published in nationally and internationally reputed journals, six book chapters, and three books. He is an editorial board and referee for several national and international journals. To date, he has supervised twenty-one MS students. Dr. Islam is currently supervising the research of two MS and six Ph.D. students in the areas of field crop production and management.

Contents

Preface **XI**

Chapter 1 **1**
Advances in Breeding in Vegetable *Brassica rapa* Crops
by María Elena Cartea, Fernando Cámara-Martos, Sara Obregón, Francisco Rubén Badenes-Pérez and Antonio De Haro

Chapter 2 **19**
Rapeseed-Mustard Breeding in India: Scenario, Achievements and Research Needs
by Subhash Chand, Om Prakash Patidar, Rajat Chaudhary, Ranjit Saroj, Kailash Chandra, Vijay Kamal Meena, Omkar M. Limbalkar, Manoj Kumar Patel, Priya P. Pardeshi and Prashant Vasisth

Chapter 3 **41**
Innovative Strategies to Develop Abiotic and Biotic Stress Tolerance in Mustard (Brassicaceae)
by Bahaderjeet Singh, Amanpreet Singh Sran and Gagandeep Singh Sohi

Chapter 4 **55**
Embryo Culture and Embryo Rescue in *Brassica*
by Mohammad Akmal

Chapter 5 **69**
Breeding Mustard (*Brassica juncea*) for Salt Tolerance: Problems and Prospects
by Jogendra Singh, Parbodh Chander Sharma and Vijayata Singh

Chapter 6 **85**
Salinity Tolerance in Canola: Insights from Proteomic Studies
by Ali Bandehagh, Zahra Dehghanian, Robert Henry and Mohammad Anwar Hossain

Chapter 7 **107**
Epidemiology, Genetics and Resistance of *Alternaria* Blight in Oilseed *Brassica*
by Subroto Das Jyoti, Naima Sultana, Lutful Hassan and Arif Hasan Khan Robin

Chapter 8 127
Breeding for Disease Resistance in Brassica Vegetables Using DNA Marker Selection
by Mst Arjina Akter, Hasan Mehraj, Takeru Itabashi, Tomoe Shindo, Masaaki Osaka, Ayasha Akter, Naomi Miyaji, Naoki Chiba, Junji Miyazaki and Ryo Fujimoto

Chapter 9 143
Brassica-Aphid Interaction: Modulated Challenges and Sustainable Approach for Management
by S.A. Dwivedi, Lelika Nameirakpam and Ajay Tomer

Preface

Brassica exists in different forms, for example, oilseed, rutabagas, vegetables, and fodder. It belongs to the family Brassicaceae (formerly Cruciferae) and is an outstanding source of oilseed crop in the world. It contributes the greatest number of edible oils after soybean and plays a vital role in human health by providing the cheapest oil for the human diet. The seeds contain 35%–45% oil, which is considered the main product, while the remaining meal after oil extraction is considered a byproduct and is widely used as a high-protein source of animal feed. Apart from its culinary purposes, *Brassica* is also used in the preparation of soaps, hair oils, lubricants, medicine, paints, and as a condiment in pickles.

Climate change has significantly decreased the growth, yield, and productivity of *Brassica* spp. due to various stress factors. Thus, high-yielding, climate-resilient, and disease-resistant varieties are required to maintain as well as increase future agricultural production. Intensive conventional breeding efforts in the past few decades have increased seed yield as well as agronomic traits. Further improvement may become exhausted and stagnant based on a single breeding approach. Therefore, to ensure food security, modern breeding approaches should be explored for the development of genetically superior *Brassica* spp. cultivars suitable for a wide range of environments. Introgression of insect and disease resistance and other desirable traits into *Brassica* spp. using inter-and/or intra-specific hybridization and biotechnological and molecular techniques could be useful. Keeping all these points in mind, this book discusses the current trends of *Brassica* breeding, genetic resources and their conservation, inheritance of important traits, breeding methods, and molecular and biotechnological approaches.

This book is an important resource for many readers, researchers, and scientists, who will find this information useful for the advancement of their research towards a better understanding of *Brassica* spp. breeding programs.

A.K.M. Aminul Islam
Professor,
Faculty of Agriculture,
Department of Genetics and Plant Breeding,
Bangabandhu Sheikh Mujibur Rahman Agricultural University,
Gazipur, Bangladesh

Mohammad Anwar Hossain and A.K.M. Mominul Islam
Bangladesh Agricultural University,
Bangladesh

Chapter 1

Advances in Breeding in Vegetable *Brassica rapa* Crops

María Elena Cartea, Fernando Cámara-Martos, Sara Obregón, Francisco Rubén Badenes-Pérez and Antonio De Haro

Abstract

Brassica rapa includes oil and vegetable crops having a variety of forms, such as oilseeds, leafy vegetables and turnips. Leafy types, which are called turnip greens and turnip tops, are popular crops in NW Spain, and they represent an important part of the diet. However, their cultivation is limited in southern areas or in the Mediterranean basin, probably due to a lack of adaptation. Still, they could occupy a prominent place in the Mediterranean diet, which is based on a high consumption of fruits and vegetables. In this review, we summarize the studies on the agronomical and nutritional value of these crops when grown under Mediterranean climate conditions. Data reported here might be useful for a deeper understanding of these crops for both nutritional quality and bioaccessibility, and for selecting varieties adapted to the two abovementioned Mediterranean conditions, as well as for organic farming systems, thus contributing to the diversification of traditional Brassica vegetable production systems.

Keywords: turnip greens, turnip tops, adaptation, bioaccessibility, nutritional quality

1. Introduction

1.1 Taxonomy and diversified morphotypes

Brassica rapa (2n = 20, synonymous with *B. campestris* L.) is an economically important species belonging to the *Brassica* genus, Brassiceae tribe, from the Brassicaceae family. The *Brassica* genus includes many important crops. Among them, relationship of six species formed the model of U's triangle, with three basic diploid species, namely *B. rapa* (A genome, n = 10), *Brassica oleracea* (C genome, n = 9) and *Brassica nigra* (B genome, n = 8), which gave rise to three amphidiploid species, namely *Brassica napus* (AC genome, n = 19), *Brassica juncea* (AB genome, n = 18) and *Brassica carinata* (BC genome, n = 17).

Brassica rapa is an important oil and vegetable crop in many parts of the world, whose seeds are used for oil, and leaves, flowers, stems and roots are used as vegetables. *B. rapa* vegetables are consumed worldwide and provide a large proportion of the daily food intake in many regions of the world. Cultivation of this species for many centuries in different parts of the world has caused a large variation in the plant organs that are consumed (roots, leaves, and flower buds), which has resulted in the human selection of different morphotypes, depending on local preferences [1].

Based on their morphological appearance and on the organs used, *B. rapa* crops can be classified into two groups:

i. Vegetable types used for their tubers (=hypocotyl), leaves and flower buds, which include the *rapa* (= *rapifera* or *ruvo*) group and the leafy vegetable forms. These vegetable types belong to six groups: *rapa*, *chinensis, pekinensis*, *parachinensis*, *nipposinica*, *perviridis* and *narinosa* [2].

ii. Oleiferous types, of which canola is a specific form, having low erucic acid levels in its oil and low glucosinolate content in its meal protein.

Until recently, these groups were considered as separate species because of the wide range of variability they show and the fact that they evolved in isolation from each other.

The *oleifera B. rapa* group includes oilseed crops that are known in Europe as rapeseed or turnip rape. It is believed that European forms developed in the Mediterranean area and then they were distributed from Europe to China. In India, crops used for oil production belong to the *trilochularis* and *dichotoma* groups. Sarson and toria types belong to this group. There are three ecotypes: brown sarson, toria and yellow sarson. Out of these, brown sarson appears to be the oldest one [2]. Yellow sarson is characterized by its yellow colored seeds and self-compatibility. Many of the cultivars have 3–4 valved siliquae, and for this reason, it was named trilochularis. It is believed to have evolved from brown sarson as a mutant and has survived because of its self-compatible nature. It might have been selected by farmers for its attractive yellow-colored seeds and bigger seed size.

Vegetable *B. rapa* crops, including rapifera and leafy types, are important crops in European and Asian countries, particularly in China, Korea, and Japan. Their consumption varies widely around the world and they are consumed as raw or steamed vegetables. The largest and most diverse *B. rapa* group consists of crops belonging to the *pekinensis* type, which includes popular crops in Chinese cuisine such as pet-sai or Chinese cabbage (**Table 1**). They are characterized by having large leaves and forming heads of different shapes. Chinese cabbage, for example, is the cabbage used for preparing dishes such as sauerkraut and kimchi, the famous fermented dish favored by Koreans. Its seeds have also been used for the hot mustard favored in Chinese cuisine. Pak-choy or bok-choy (*chinensis* group) are also popular crops in Asian culture. They have been used for their leaves, which do not form heads and are smooth. It is assumed that pak-choy types with narrow or wide green-white petioles were the first *B. rapa* crops to evolve in Central China. Another group of cultivars that is characterized by many narrow leaves belong to the *perviridis* group, which includes neep greens from Europe and the Japanese cultivar Komatsuna. Finally, we have the *nipposinica* group, which includes Japanese crops like mizuna or mibuna, which can be eaten raw or cooked at any stage, from seedling to mature plant (**Table 1**).

The *rapa* or *rapifera* group is characterized by the thickening of the hypocotyls, which can show different colors and shapes, and has a mainly horticultural and forage use. Turnips are both cultivated as fodder crops or as vegetables, and depending on the region, the tubers, leaves and shoots are used. Turnip greens are the young leaves harvested in the vegetative growth period. Turnip tops are the fructiferous stems with flower buds and the surrounding leaves that are consumed before opening and while still green (**Table 1**, **Figure 1**). In Europe, they are notably popular in Portugal, Italy and Spain, where they play an important role in traditional farming and in the diet. In these countries, *B. rapa* includes two main crops, turnip greens and turnip tops, as vegetable products. They are commonly consumed as boiled

Group	Crops	Distribution	Plant part used
Vegetable types			
rapa (= rapifera)	Turnip, turnip greens, turnip tops, rapini, broccoletti di rape, brocoletto, turnip broccoli, cima di rapa, Italian turnip	Europe	Leaves, flower buds and hypocotyl
chinensis	Pak-choy, Bok-choy, celery mustard	China	Leaves
pekinensis	Chinese cabbage, napa cabbage, celery cabbage, pet-sai, napa, wong-bok, chihli	China, Korea, Taiwan, Japan	Leaves
parachinensis	Choi-sum, caixin, caitai	China	Leaves and flower buds
nipposinica	Mizuna, mibuna, curled mustard, Japanese greens	Japan	Leaves
perviridis	Komatsuna, spinach mustard, tendergreen, neep greens	Japan, Korea, Taiwan	Leaves
narinosa	Wutacai or heibaicai	China	Leaves
Oleifera types			
oleifera	Turnip rape, rapeseed	China	Seeds
dichotoma	Brown sarson, toria	India	Seeds
trilochularis	Yellow sarson	India	Seeds

Table 1.
Taxonomic groups in Brassica rapa *species.*

Figure 1.
Leafy vegetable crops from the Brassica rapa *group: Turnip (A), turnip greens (B) and turnip tops (C).*

vegetables, being used in the preparation of soups and stews and they have a slightly spicy flavor like mustard greens [3]. Turnip greens and turnip tops have good commercial prospects in both countries and, the number of companies selling *B. rapa* canned products has been increasing in the last years.

1.2 Origin of *Brassica rapa* crops

The origin of cultivated *B. rapa* crops is still unknown. This species was probably the first domesticated Brassica several millennia ago, as a multipurpose crop [4]. It is believed that the most likely explanation for the wide variation within this species is that cultivated forms arose independently in different places of the world from wild *B. rapa* [1]. It seems to have spread naturally to the Western

Mediterranean region and to Central Asia, with secondary centers of diversity in Europe, Western Russia, Central Asia, and the Near East [5].

According to the studies based on morphology, geographic distribution, isozymes and molecular data, cultivated subspecies of *B. rapa* most likely originated independently in two different centers—Europe and Asia. Europe should be one primary center of origin for oil and turnip types [4], whereas East Asia should be another primary center for Indian oil types and leafy vegetables [1, 6]. Today, it is well established that Asia represents the main area of diversification for vegetable *B. rapa* crops. Leafy vegetables such as Chinese cabbage, pak-choi and narinosa may have been first domesticated in China. China is also the center of origin of Chinese turnip rape (var. *oleifera*). Other accessions of *B. rapa* most likely derived from different morphotypes in the two centers of origin and subsequently evolved separately.

It is believed that *B. rapa* was introduced into China through Western Asia or Mongolia as an agricultural species. In fact, *B. rapa* is also recognized as the ancestor of many oriental Brassica vegetables. Its introduction into Japan could have occurred via China or Siberia. In India, *B. rapa* is cultivated as an oilseed, but no wild forms are known in this country. In East Asia, leafy types such as Chinese cabbage, bok choy, pak-choi, mizuna, celery mustard, and Chinese kale, among others, are used extensively as vegetables [6]. In China, flowers of the crop called choy-sum (*parachinensis* group) are also consumed, and these inflorescences are known as caixin or caitai.

In Europe, broccoleto types, turnip rape and turnips are the predominant forms [7] and they can be used for both as food and feed. Other *B. rapa* accessions most likely derived from different morphotypes in the two centers of origin and subsequently evolved separately. The *rapa* or *rapifera* group is believed to have evolved in Europe. It is supposed that it was first used for its nutritious root around 2,500–2,000 B.C. and spread to other parts of the world afterwards. The expansion of vegetable crops within this group such as turnip greens and turnip tops took place later on and independently from the origin of leafy forms in Asia [7].

1.3 Breeding for turnip greens and turnip tops

This review will be focused on two *B. rapa* crops: turnip greens and turnip tops. In Northwestern Spain, Portugal and Southern Italy, both crops have a long tradition and they represent two important commodities, being part of very traditional recipes. Like other Brassica vegetable crops, they are generally either eaten after being cooked or they can also be processed as canned foods. Turnip greens and turnip tops have good commercial prospects and their consumption, both fresh and processed, has increased considerably in the last years. New uses and new markets for these crops (canned, frozen, fourth range-foods, ...) have been grown lately.

A collection of local varieties of turnip greens and turnip tops from Northwestern Spain is currently kept at the Misión Biológica de Galicia (CSIC) in Pontevedra, Northwestern Spain. These landraces are a valuable resource, since they are adapted to the climatic conditions of this area. Agronomical and nutritional evaluations of this collection were previously performed by [8–10]. Authors reported a high genetic diversity for several agronomic traits and found that some varieties are a valuable source of bioactive compounds such as glucosinolates and phenolic compounds. However, their cultivation is limited in southern areas or in the Mediterranean basin, probably due to a lack of adaptation. Still, these crops could occupy a prominent place in the Mediterranean diet, which is based on a high consumption of fruits and vegetables. The evaluation of *B. rapa* varieties with wide adaptability across diverse farming environments becomes essential for selecting varieties for future breeding programs based on producers' and consumers' preferences. With this goal in mind, a breeding program in turnip tops and turnip greens was started at IAS-CSIC in Córdoba (South of Spain) in recent years. The goal was to achieve varieties adapted to the environmental conditions of this area but preserving similar nutritional properties to those produced in their original region.

In this review, we summarize the studies on the agronomical and nutritional value of these crops grown under Mediterranean climate conditions. Data reported here might be useful for deeper understanding of these crops for both nutritional quality and bioaccessibility, resistance to biotic stress, and for selecting varieties adapted to Mediterranean conditions, thus contributing to the diversification of traditional Brassica vegetable production systems.

2. Characterization, evaluation and selection of *Brassica rapa* germplasm under Mediterranean conditions

2.1 Introduction

It is well known that the change from a Western dietary pattern (high consumption of calories, animal products and sugars) to a Mediterranean diet (high consumption of fruits, vegetables, grains, legumes and reduced amounts of animal products, with the use of olive oil as the preferred fat) reduces the risk of diabetes by 7%, heart disease by 10%, and total mortality by 8% [11, 12]. It is therefore clear that increasing and promoting the consumption of locally produced foods of plant origin (km. 0), and, in particular, those with nutraceutical properties, is one of the crucial factors for the well-being and health promotion, hence allowing the prevention of various diseases, such as cancer, and cardiovascular and neurodegenerative diseases [13].

For several years now, the IAS-CSIC research group in plant breeding has been studying the possibilities of producing turnip greens and turnip tops in Southern Spain for incorporation into the Mediterranean diet. Under these conditions, they could be considered as a new regional crop that provides vegetables with more interesting nutraceutical and organoleptic properties than Brassica species, such as cauliflower, broccoli or Brussels sprouts, which have seen their consumption reduced mainly in children due to their strong and peculiar smell and taste.

The goal of this work was to study the adaptation and cultivation of a collection of germplasm and cultivars of *B. rapa* harvested in Galicia (Northwestern Spain) in the Guadalquivir Valley, and select the lines with better agronomic and nutritional characteristics, hence expanding the usual consumption area. In the evaluation and selection process, the turnip greens and turnip tops production capacity, as well as the glucosinolate content of the harvested products as a quality criterion for the final product, were studied.

2.2 Plant material

The *B. rapa* L. var. *rapa* germplasm used for this work came from the Brassica Germplasm Bank at Misión Biológica de Galicia (Pontevedra, Northwestern Spain), where it had been characterized by its agronomic characteristics and its aptitude for turnip greens and turnip tops production.

2.3 First trials of *Brassica rapa* cultivation in Córdoba (2009-2012)

The effect of different sowing dates on the production and quality of turnip greens and turnip tops was studied during the first stage. For this purpose, five

B. rapa accessions from the MBG-CSIC Germplasm Bank selected by their differences in phenological growth cycle (early and late) were used. These five accessions were cultivated in Córdoba (Southern Spain, Guadalquivir Valley) during the 2009/10, 2010/11 and 2011/12 agricultural seasons. Different sowing dates were tested for each agricultural season in order to cover the largest potential period of turnip greens and turnip tops production. During the 2009/10 season, the same entries were grown in Pontevedra, being used as a control.

During the first season that the five *B. rapa* accessions were sown in Cordoba (2009/10), the low turnip greens production for all entries highlighted the inadequacy of the sowing dates chosen and the need to bring them forward in successive seasons. In the first sowing, turnip greens production was low and turnip tops of acceptable quality were not obtained. The second sowing was lost due to the unusually high rainfall that caused root asphyxiation and plant death. In the third sowing, a good turnip greens production was achieved but the increase in spring temperatures caused them to rise quickly, thus obtaining low-quality turnip tops.

These results determined that all sowing dates would have to be brought forward in the following seasons (2010/11 and 2011/12), starting in September. This change notably favored the crop adaptation in Córdoba, improved plant development in the field and improved the turnip greens and turnip tops production. The existence of accessions, that did not form quality turnip tops in Córdoba, revealed the need to extend the germplasm collection to be studied, in order to be able to select the most suitable genotypes for turnip tops production in Mediterranean edaphoclimatic conditions (**Table 2**).

2.4 Characterization of a *Brassica rapa* germplasm collection (2013–2014)

Once the optimal sowing date was adjusted, in the next stage (2013/14 agricultural season), characterization and evaluation of 19 *B. rapa* accessions also from the MBG-CSIC Germplasm Bank was carried out. The selection of these entries was made according to the agronomic characteristics and phenological cycle in their origin area. A randomized block design with 3 replications was used in all trials. Glucosinolate analysis of was carried out in accordance with the European standard for this determination [14].

Location	Season	Transplanting date	Turnip greens harvest	Turnip tops harvest
Pontevedra (Control)	2009/10	September	December	January to April
Córdoba	2009/10	January	April	No
		March	No	No
		April	June	July
	2010/11	September	December	January
		November	April	April
		January	No	No
	2011/12	September	November	December
		November	February	February
		January	No	No

Table 2.
Transplanting and harvesting dates of Brassica rapa *accessions in each localition by season and sowing date.*

The cultivation of these entries in Córdoba was successful, and turnip greens and turnip tops harvest was abundant for almost all the entries (**Figure 2**).

In addition to the agronomic evaluation, the glucosinolate content of the turnip greens and turnip tops harvested for each of the entries was analyzed. In general, the average glucosinolate content of turnip greens (27.98 μmol/g dry matter) was lower than that of turnip tops (30.25 μmol/g dry matter), which highlights the high variability in glucosinolate content between the different accessions and within each accession. The glucosinolate pattern was similar in turnip greens and turnip tops, with gluconapin being the major glucosinolate (representing about 80% of total glucosinolates), followed by progoitrin, glucobrassicanapin, gluconapoleiferin, glucobrassicin, 4-metoxiglucobrassicin and neoglucobrassicin. Similar results were found in previous works on the glucosinolate content in vegetable *B. rapa* crops [10, 15]. No differences were found between the glucosinolate profile of the samples collected in Córdoba and that of samples collected in Pontevedra. Some accessions cultivated in Córdoba stood out for their ability to produce turnip greens and/or turnip tops with a total glucosinolate content equal or greater than those produced at their usual cultivation place (**Figure 3**).

2.5 Evaluation of selected *Brassica rapa* accessions

In the third stage (2014/15 season) six accessions were cultivated in Córdoba (from the 19 studied in the previous season), which were selected based on their homogeneity and their turnip tops production in Córdoba. The entries chosen were evaluated in terms of their agronomic characteristics, productivity and glucosinolate content of the harvested turnip greens and turnip tops.

Agronomic evaluations were carried out throughout the entire cultivation cycle in Córdoba, and it was possible to harvest quality turnip greens and turnip tops in all cultivated accessions (**Figure 4**). In general, we obtained turnip greens in Córdoba with lower fresh weight than that of turnip greens produced in Pontevedra. The opposite occurred with the fresh weight of turnip tops and the number of turnip tops/plant, which was higher in the entries cultivated in Córdoba (**Table 3**).

Figure 2.
Brassica rapa *cultivation at the IAS-CSIC experimental farm, Córdoba (season 2013–2014).*

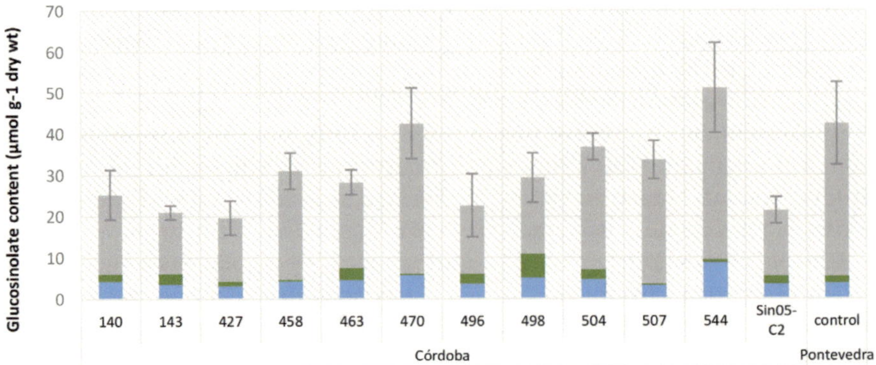

	140	143	427	458	463	470	496	498	504	507	544	Sin05-C2	control
mean GNA	19.29	14.85	15.36	26.33	20.78	36.32	16.62	18.58	29.76	29.99	41.70	15.90	37.12
mean PRO	1.68	2.59	1.12	0.42	2.90	0.44	2.28	5.70	2.36	0.42	0.74	1.97	1.62
mean others	4.35	3.56	3.20	4.32	4.62	5.77	3.73	5.14	4.71	3.24	8.70	3.59	3.78

Figure 3.
Gluconapin (GNA), progoitrin (PRO) and other glucosinolate contents (mean ± standard deviation) in turnip tops of Brassica rapa *accessions cultivated in Córdoba and Pontevedra. Error bars represent the standard deviation of total glucosinolates.*

Figure 4.
Samples of turnip tops harvested in Córdoba (season 2014–2015).

	Turnip greens		Turnip tops			
	FW[a] (g)	M[b] (%)	Fw (g)	M (%)	Stems (n°)	T[c]
BRS0143	6.04	80.94	49.46	90.85	20.58	120
BRS0427	5.37	80.04	47.14	90.27	17.80	98
BRS0496	5.85	80.69	33.44	90.33	12.14	98
BRS0498	7.33	79.09	113.87	92.88	19.94	134
BRS0504	7.55	81.80	107.11	93.09	19.39	134
BRSin05-C2	5.73	79.70	45.36	85.58	21.84	106
Mean Cordoba	6.43	80.38	66.06	90.50	18.62	115
*Mean Pontevedra	22.12	90	63.02	91	12.13	162.3

Source: Francisco et al., [9].
[a]*FW: fresh weight.*
[b]*M: moisture.*
[c]*D: days from turnip tops sowing to harvest.*

Table 3.
Agronomic characteristics of turnip greens and turnip tops harvested in Córdoba (season 2014–2015).

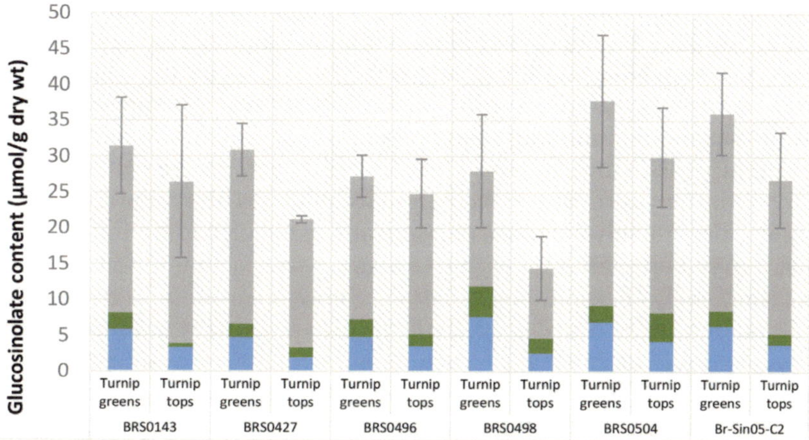

	BRS0143 Turnip greens	BRS0143 Turnip tops	BRS0427 Turnip greens	BRS0427 Turnip tops	BRS0496 Turnip greens	BRS0496 Turnip tops	BRS0498 Turnip greens	BRS0498 Turnip tops	BRS0504 Turnip greens	BRS0504 Turnip tops	Br-Sin05-C2 Turnip greens	Br-Sin05-C2 Turnip tops
average GNA	23.21	22.52	24.24	17.83	19.95	19.57	16.05	9.76	28.53	21.64	27.51	21.51
average PRO	2.30	0.50	1.85	1.38	2.46	1.66	4.28	2.14	2.36	4.05	2.10	1.50
average other	5.92	3.44	4.82	2.00	4.84	3.62	7.66	2.59	6.95	4.26	6.43	3.82

Figure 5.
Gluconapin (GNA), progoitrin (PRO) and other glucosinolate contents (mean ± standard deviation) in turnip greens and turnip tops of Brassica rapa *selected accessions cultivated in Córdoba (season 2014–2015).*

The total glucosinolate content was significantly higher in turnip greens than in turnip tops. Accessions BRS0143, BRS0504 and BRSin05-C2 stood out for their capacity to produce turnip greens and turnip tops with high gluconapin content (**Figure 5**). There are numerous studies that indicate that gluconapin is beneficial for health, since its degradation product (3-butenyl isothiocyanate) is capable of producing cell death induced mainly through tumor cell necrosis [16–18]. These results indicate the potential of these selected accessions to obtain varieties that are capable of producing turnip greens and turnip tops with high levels of beneficial glucosinolates (gluconapin) and low levels of glucosinolates with anti-nutritional potential (progoitrin).

3. Glucosinolate bioaccessibility

Bioavailability can be defined as being the micronutrient or bioactive compound fraction, originally present in the food, which is solubilized and absorbed in the intestinal lumen, metabolized by typical routes, and finally used for typical physiological functions or deposited in storage compounds [19]. As glucosinolates are hydrolyzed by the enzyme myrosinase, into glucose and a wide variety of unstable aglycones such as isothiocyanates, thiocyanates, nitriles, indoles, thiones and epithioalkanes among others, knowing the beneficial physiological effect of all these compounds requires a wide variety of further in vivo studies.

A first step could be to focus on the amount of glucosinolates that come into contact with enterocytes. Thus, bioavailability studies can be partly replaced by bioaccessibility ones. This term refers to the fraction of the micronutrient or bioactive compound that is soluble in the intestinal lumen and therefore will be capable of being absorbed by the enterocytes of the small intestine [20]. Bioaccessibility studies are based on a simulated gastrointestinal food digestion formed by an oral phase with salivary amylase, a gastric phase with pepsin-HCl at pH 2, and later by an intestinal phase with pancreatin-bile salts [21, 22]. Finally, the digest is centrifuged and glucosinolate fraction is determined, as the amount of this compound present in the supernatant.

Several studies have shown that around 85% of the initial glucosinolate dose in a rapeseed meal is capable of resisting the physiological conditions of the stomach, and around 63–75% remains intact after *in vitro* simulation of a 4 h digestion in the small intestine [23]. Another study [24], using simulated *ex vivo* gastrointestinal digestion, also gave bioaccessibility values of 71 and 29% for two glucosinolates (glucoraphenin and glucoraphasatin) of *Matthiola incana.* The presence of the sulphate group and thioglucose moiety confers the glucosinolate molecule with high water solubility [23]. However, this bioaccessibility percentage will depend on the structure of the glucosinolate molecule and its ability to bind non-specifically to macromolecules (mainly proteins, peptides and small glycoproteins).

Thus, a previous study [25] with five plant species belonging to the Brassicaceae family (*B. rapa*, *B. oleracea*, *B. carinata*, *E. vesicaria* and *S. alba*) showed that over 30% of the glucosinolates initially present in the leaves of this plant species would be capable of reaching human enterocytes, hence resisting the degradation processes of digestive enzymes, including its own myrosinase enzyme (**Figure 6**). In that study, the highest bioaccessibility percentages corresponded to indolic glucosinolates such as glucobrassicin (70%) and neoglucobrassicin (around 56%), followed by aliphatic ones such as progoitrin (49%) and sinigrin (32–43%). The lowest bioaccessibility percentages corresponded to aromatic glucosinolates, with a percentage of 25% for sinalbin.

Another similar study conducted by [26] also showed the highest percentage of bioaccessibility for an indolic glucosinolate like glucobrassicin (around 42%) in broccoli "Parthenon" and Savoy cabbage "Dama" brassicas. According to these results, the presence of a five-membered pyrrole ring fused to a benzene ring seems to confer the glucosinolate molecule a higher solubility and less uptake to other molecules from the enzymatic digestion of food than the aromatic glucosinolate group.

It is suggested that intact glucosinolates must pass through the gut epithelium by passive, facilitated or active transport [23], although the real path way remains unknown. It is also important to emphasize that several glucosinolate hydrolysis-derived products, such as isothiocyanates and indoles, can also be found in the small intestine, likely arising by the enzymatic processing mediated by the plant

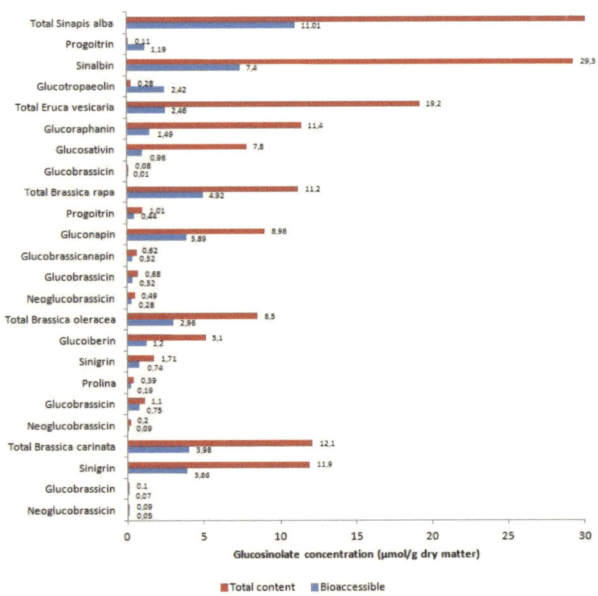

Figure 6.
Glucosinolate concentration (total and bioaccessible) in five plant species belonging to the Brassicaceae *family (expressed as µmol/g dw).*

myrosinase. Bioaccessibility studies should also include these compounds because both glucosinolates and their derivatives provide beneficial effects on human health. Thus, studies with radioisotopes in rats [27, 28] have shown a high absorption of isothiocyanates, with a blood peak observed 3 h after ingestion. De la Fuente et al. [29] have reported bioaccessibility percentages ranging between 31 and 63% for total isothiocyanates of Brassica microgreens. Among all the isothiocyanates, one of the most studied is sulforaphane, which is produced by the hydrolysis of the glucosinolate glucoraphanin present in broccoli. A bioaccessibility study conducted by [30] has shown a concentration for sulforaphane and sulforaphane nitrile of 10.4 and 49.9 μmol/100 g of fresh broccoli after the gastric phase and 28.6 and 113 μmol/100 g of fresh broccoli after the intestinal phase. However, there is a wide variety of isothiocyanates coming from enzymatic hydrolysis of other glucosinolates whose bioavailability has not been studied yet. More research is needed in this field in order to know the nutritional role of all these compounds.

Finally, glucosinolates that are not absorbed in the small intestine reach the colon, where they could be hydrolyzed with bacterial myrosinase in nitriles and other unspecified products [31]. Formation of products from glucosinolates by intestinal microbiota is also still poorly documented and further studies are equally necessary.

4. Insect pests and diseases

Among the insect pests affecting *B. rapa* crops, the diamondback moth, *Plutella xylostella* L. (Lepidoptera: Plutellidae) and cabbage root flies, *Delia* spp. (Diptera: Anthomyiidae) are considered the most damaging pests [32–34]. Other important insect pests include *Phyllotreta* spp. (Coleoptera: Chrysomelidae) flea beetles, cabbage aphid, *Brevicoryne brassicae* L. (Hemiptera: Aphididae), cabbage butterflies, *Pieris* spp. (Lepidopera: Pieridae), cabbage moth, *Mamestra brassicae* L. (Lepidoptera: Noctuidae), pollen beetle, *Meliegethes aeneus* F. (Coleoptera: Nitidulidae), cabbage seed pod weevil, *Ceutorhynchus obstrictus* Marsham (Coleoptera: Curculionidae) [35–37]. At present, turnip greens/tops resistant varieties to major pests are scarce and chemical control is the most used method to protect these crops. Because of its attractiveness to insects, *B. rapa* has also been proposed as a trap crop and insectary plant [38].

The role of glucosinolates on pest resistance has been extensively studied in Brassica crops. Glucosinolates are considered a source of resistance to the cabbage moth, *M. brassicae*, and to the specialist *Pieris rapae* [39]. The yellow flowers of *B. rapa* are very attractive to pollen beetle, *Meligethes aeneus*, and the glucosinolate content in the inflorescence is positively correlated with *M. aeneus* incidence [40]. The content of certain glucosinolates is associated to an increased developmental time and reduced weight in the cabbage seedpod weevil, *Ceutorhynchus obstrictus* [41]. Since glucosinolate content can increase susceptibility to *P. xylostella*, breeding programs leading to increased glucosinolate content can result in higher damage by this insect [42]. Varieties with less wax on their leaves can be partly resistant to *P. xylostella* and *B. brassicae* damage [43]. However, an increase in leaf epicuticular waxes diminishes plant damage by *Phyllotreta* spp. [43].

Although *B. rapa* tends to be quite susceptible to *D. radicum*, some turnip greens/tops accessions have been identified by our group at MBG-CSIC, as they show some resistance to this pest [32]. We noticed that direct damage, as a result of *D. radicum* larvae feeding on root tissue, and indirect damage, by facilitating the entry of secondary root pathogens, reduce both yield and quality of these vegetables and eventually induce plant death (**Figure 7**).

The main diseases affecting *B. rapa* crops include fungal, bacterial and viral diseases. The most important are downy mildew (*Hyaloperonospora parasitica* (Pers.) Constant.), Turnip mosaic virus (TuMV), clubroot (*Plasmodiophora brassicae* Woronin), and soft rot caused by the bacterium *Pectobacterium carotovorum* (Jones) Waldee (syn. *Erwinia carotovora*) and *Pseudomonas marginalis* (Brown) Stevens, black rot (*Xanthomonas campestris* pv. *campestris* (Pammel) Dowson), (Xcc) and Fusarium wilt (*Fusarium oxysporum* f. sp. conglutinans/rapae) [44].

Among these, black rot of crucifers caused by Xcc is considered one of the most important diseases affecting crucifers worldwide. It is particularly destructive to *B. oleracea* vegetables because it causes reduction in yield and quality but it can also attack all other *Brassica* spp. In *B. rapa*, the disease has been reported in Chinese cabbage and other oriental *B. rapa* vegetable crops, and it can also be serious in turnip and turnip greens [45] (**Figure 8**).

A variety of resistance genes and QTLs to different diseases have been identified to develop disease resistance in *B. rapa* [44, 46]. The role of glucosinolate content against Brassica-pathogenic bacteria and fungi has been also reported from *in vitro* and *in vivo* studies, supporting the fact that they can be used as a means of disease resistance [47, 48].

Figure 7.
*Aspect of turnip greens damaged by cabbage maggot (*Delia radicum*) larvae (left). Turnip tops plants died by the attack of* Delia radicum *under natural infestation. Plants show the most common feeding symptoms with plant yellowing, stunting and slow growth (right).*

Figure 8.
Black rot, caused by bacterium Xanthomonas campestris *pv.* campestris *(Pammel) Dowson (Xcc), is considered one of the most serious diseases for crucifers worldwide. The pathogen produces V shaped necrotic lesions from leaf margins, which decrease the quality of product quality for fresh-market sale and cause a decrease in the quality trade for the food industry.*

5. Conclusions

In summary, the field and laboratory work carried out at the Institute of Sustainable Agriculture (Cordoba) in collaboration with the Misión Biológica de Galicia from 2009 to date has demonstrated the possibility of producing turnip greens and turnip tops in the Guadalquivir Valley with a performance and quality similar to those of the traditional farming area. The screening and evaluation of a collection of germplasm from the Misión Biológica de Galicia has allowed us to select the most suitable entries to obtain turnip tops with high glucosinolate content, which are beneficial to health and have organoleptic properties similar to those harvested in Galicia. The introduction of *B. rapa* cultivation in Andalusia and other similar regions would increase the diversification of horticultural products and stimulate the consumption of healthy products among the Spanish population. Data reported here might be useful for at deeper understanding of these crops for both nutritional quality and bioaccessibility, resistance to biotic stress, and for selecting varieties adapted to the Mediterranean conditions mentioned in this work.

Acknowledgements

This research was financially supported by project RTI2018-096591-B-I00 34 (MCIU/AEI/FEDER, UE). We acknowledge support of the publication fee by the CSIC Open Access Publication Support Initiative through its Unit of Information Resources for Research (URICI)-CSIC.

Conflict of interest

The authors declare no conflict of interest.

Author details

María Elena Cartea[1]*, Fernando Cámara-Martos[2], Sara Obregón[3], Francisco Rubén Badenes-Pérez[4] and Antonio De Haro[3]

1 Group of Genetics, Breeding and Biochemistry of Brassica, Misión Biológica de Galicia (MBG – CSIC), 36143, Pontevedra, Spain

2 Department of Food Science and Technology, University of Córdoba, Rabanales Campus, Building C-1, 14014, Córdoba, Spain

3 Department of Plant Breeding, Instituto de Agricultura Sostenible (IAS – CSIC), 14004, Córdoba, Spain

4 Department of Plant Protection, Instituto de Ciencias Agrarias (ICA – CSIC), 28006, Madrid, Spain

*Address all correspondence to: ecartea@mbg.csic.es

References

[1] Zhao J, Wang X, Deng B, Lou P, Wu J, Sun R, Xu Z, Vromans J, Koornneef M, Bonnema G. Genetic relationships within *Brassica rapa* as inferred from AFLP fingerprints. Theoretical and Applied Genetics. 2005;110:1301-1314. DOI: 10.1007/s00122-005-1967-y

[2] Sun R. Economic/academic importance of *Brassica rapa*. In: Xiaowu W, Chittaranjan K, editors. The *Brassica rapa* Genome. 1st ed. New York: Springer; 2015. p. 1-15. DOI: 10.1007/978-3-662-47901-8_1

[3] Padilla G, Cartea ME, Rodríguez VM, Ordás A. Genetic diversity in a germplasm collection of *Brassica rapa subsp rapa* L. from northwestern Spain. Euphytica. 2005;145:171-180. DOI: 10.1007/s10681-005-0895-x

[4] Gómez-Campo C. Taxonomy. In: Gómez-Campo C, editor. Biology of *Brassica* coenospecies. 1st ed. Amsterdam, Holland: Elsevier Science BV; 1999. p. 2-32.

[5] Tsunoda S. Eco-physiology of wild and cultivated forms in Brassica and allied genera. In: Tsunoda S, Hinata K, Gómez-Campo C, editors. Brassica crops and wild allies. Biology and breeding. 1st ed. Tokio, Japan: Japan Scientific Societies Press; 1980. p. 109-119.

[6] Guo Y, Chen S, Li Z, Cowling WA. Center of origin and centers of diversity in an ancient crop, *Brassica rapa* (turnip rape). Journal of Heredity. 2014;105:555-565. DOI: 10.1093/jhered/esu021

[7] Gómez-Campo C, Prakash S. Origin and domestication. In: Gómez-Campo C, editor. Biology of *Brassica* coenospecies. 1st ed. Amsterdam, Holland: Elsevier Science B.V; 1999. p. 33-52.

[8] Padilla G, Cartea ME, Velasco P, de Haro A, Ordás A. Variation of glucosinolates in vegetable crops of *Brassica rapa*. Phytochemistry. 2007;68:536-545. DOI: 10.1016/j.phytochem.2006.11.017

[9] Francisco M, Velasco P, Lema M, Cartea ME. Genotypic and environmental effects on agronomic and nutritional value of *Brassica rapa*. Agronomy Journal. 2011;103:735-742. DOI: 10.2134/agronj2010.0439

[10] Cartea ME, de Haro A, Obregón S, Soengas P, Velasco P. Glucosinolate variation in leaves of *Brassica rapa* crops. Plant Foods for Human Nutrition. 2012;67:283-288. DOI: 10.1007/s11130-012-0300-6

[11] Dinu M, Pagliai G, Casini A, Sofi F. Mediterranean diet and multiple health outcomes: An umbrella review of meta-analyses of observational studies and randomised trials. European Journal of Clinical Nutrition. 2018;72:30-43. DOI: 10.1038/ejcn.2017.58

[12] Satija A, Bhupathiraju SN, Rimm EB, Spiegelman D, Chiuve SE, Borgi L, et al. Plant-based dietary patterns and incidence of type 2 diabetes in US men and women: Results from three prospective cohort studies. PLoS Medical. 2016;13:e1002039. DOI: 10.1371/journal.pmed.1002039

[13] Lappé FM. La dieta ecológica: cómo cocinar, sin carnes, platos de elevado nivel nutritivo. 1 st ed. Barcelona: Roselló Impressions DL; 1987. 220 p.

[14] ISO9167-1. Determination of glucosinolates content-Part I: Method using high-performance liquid chromatography. ISO 9167-1:1992. International Stand Organ Geneva, Switz. 1992;1-9.

[15] Hanson P, Yang R, Chang L, Ledesma L, Ledesma D. Carotenoids,

ascorbic acid, minerals, and total glucosinolates in choysum (*Brassica rapa* cv. *parachinensis*) and kailaan (*B. oleracea alboglabra* group) as affected by variety and wet and dry season production. Journal of Food Composition and Analysis. 2011;24:950-962. DOI: 10.1016/j.jfca.2011.02.001

[16] Mithen RF, Dekker M, Verkerk R, Rabot S, Johnson IT. The nutritional significance, biosynthesis and bioavailability of glucosinolates in human foods. Journal of the Science of Food and Agriculture. 2000;80:967-984. DOI.org/10.1002/(SICI)1097-0010

[17] Kadir NHA, David R, Rossiter JT, Gooderham NJ. The selective cytotoxicity of the alkenyl glucosinolate hydrolysis products and their presence in Brassica vegetables. Toxicology. 2015;334:59-71. doi.org/10.1016/j.tox.2015.06.002

[18] Obregón S. Estudio del contenido y valor nutracéutico de los glucosinolatos y otros compuestos presentes en nabizas y grelos (*Brassica rapa* L. var. *rapa*) cultivados en el sur de España. [thesis]. Cordoba University; 2016

[19] Cámara-Martos F, Pérez-Rodríguez F, Amaro-López MA, Moreno-Rojas R. Zinc as essential micronutrient: Physiological role and factors affecting its bioavailability. In: Betancourt AI, Gaitán HF, editors. Micronutrients: sources, properties and health effects. 1st ed. New York: Nova Biomedical; 2011; p. 37-64.

[20] Ramírez-Ojeda AM, Moreno-Rojas R, Sevillano-Morales J, Cámara-Martos F. Influence of dietary components on minerals and trace elements bioaccessible fraction in organic weaning food: A probabilistic assessment. European Food Research and Technology. 2017;243:639-650. https://doi.org/10.1007/s00217-016-2777-y

[21] Egger L, Ménard O, Delgado-Andrade C, Alvito P, Assunção R, Balance S, et al. The harmonized INFOGEST *in vitro* digestion method: From knowledge to action. Food Research International. 2016;88:217-225. https://doi.org/10.1016/j.foodres.2015.12.006

[22] Minekus M, Alminger M, Alvito P, Ballance S, Bohn T, Bourlieu C, et al. A standardised static in vitro digestion method suitable for food–an international consensus. Food and Function. 2014;5:1113-1124. DOI: 10.1039/c3fo60702j

[23] Holst B, Williamson G. A critical review of the bioavailability of glucosinolates and related compounds. Natural Products Reports. 2004;21:425-447. DOI: 10.1039/b204039p

[24] Blažević I, Đulović A, Burčul F, Popović M, Montaut S, Bilušić T, Vrca I, Markić J, Ljubenkov I, Ruščić M, Rollin P. Stability and bioaccessibility during ex vivo digestion of glucoraphenin and glucoraphasatin from *Matthiola incana* (L.) R. Br. Journal of Food Composition and Analysis. 2020;103483. https://doi.org/10.1016/j.jfca.2020.103483

[25] Cámara-Martos F, Obregón-Cano S, Mesa-Plata O, Cartea-González ME, de Haro-Bailón A. Quantification and in vitro bioaccessibility of glucosinolates and trace elements in Brassicaceae leafy vegetables. Food Chemistry. 2021;339:127860. https://doi.org/10.1016/j.foodchem.2020.127860

[26] Fernández-León AM, Fernández-León MF, González-Gómez D, Ayuso MC, Bernalte MJ. Quantification and bioaccessibility of intact glucosinolates in broccoli 'Parthenon'and Savoy cabbage "Dama". Journal of Food Composition and Analysis. 2017;61:40-46. https://doi.org/10.1016/j.jfca.2016.11.010

[27] Bollard M, Stribbling S, Mitchell S, Caldwell J. The disposition of allyl isothiocyanate in the rat and mouse. Food and Chemical Toxicology. 1997;35:933-943. https://doi.org/10.1016/S0278-6915(97)00103-8

[28] Conaway CC, Jiao D, Kohri T, Liebes L, Chung FL. Disposition and pharmacokinetics of phenethyl isothiocyanate and 6-phenylhexyl isothiocyanate in F344 rats. Drug Metabolism and Disposition. 1999;27:13-20.

[29] De la Fuente B, López-García G, Máñez V, Alegría A, Barberá R, Cilla A. Evaluation of the bioaccessibility of antioxidant bioactive compounds and minerals of four genotypes of Brassicaceae microgreens. Foods. 2019;8:250. DOI: 10.3390/foods8070250

[30] Sarvan I, Kramer E, Bouwmeester H, Dekker M, Verkerk R. Sulforaphane formation and bioaccessibility are more affected by steaming time than meal composition during in vitro digestion of broccoli. Food Chemistry. 2017;214:580-586. https://doi.org/10.1016/j.foodchem.2016.07.111

[31] Barba FJ, Nikmaram N, Roohinejad S, Khelfa A, Zhu Z, Koubaa M. Bioavailability of Glucosinolates and their breakdown products: Impact of Processing. Frontiers in Nutrition. 2016; 3:24. DOI: 10.3389/fnut.2016.00024

[32] Santolamazza-Carbone S, Velasco P, Cartea ME. Resistance to the cabbage root fly, *Delia radicum* (Diptera, Anthomyiidae), of turnip varieties (*Brassica rapa subsp. rapa*). Euphytica. 2017; 213:274. https://doi.org/10.1007/s10681-017-2069-z

[33] Furlong MJ, Wright DJ, Dosdall LM. Diamondback moth ecology and management: problems, progress, and prospects. Annual Review of Entomology. 2013;58:517-541. DOI: 10.1146/annurev-ento-120811-153605

[34] Soroka JJ, Weiss RM, Grenkow LF, Olfert OO. Relationships among root maggot (Delia spp., Diptera: Anthomyiidae) infestation, root injury, and seed yields of canola *Brassica napus* L. and *Brassica rapa* L. Canadian Journal of Plant Science. 2020;100:575-591. DOI:10.1139/cjps-2019-0278

[35] Badenes-Perez FR, Shelton AM. Pest management and other agricultural practices among farmers growing cruciferous vegetables in the Central and Western highlands of Kenya and the Western Himalayas of India. International Journal of Pest Management. 2006;52:303-315. DOI: 10.1080/09670870600819169

[36] Cárcamo H, Olfert O, Dosdall L, Herle C, Beres B, Soroka J. Resistance to cabbage seedpod weevil among selected Brassicaceae germplasm. Canadian Entomologist. 2007;139:658-669. DOI: https://doi.org/10.4039/n06-083

[37] Cartea ME, Padilla G, Vilar M, Velasco P. Incidence of the major Brassica pests in Northwestern Spain. Journal of Economy Entomology. 2009;102:767-773. https://doi.org/10.1603/029.102.0238

[38] Badenes-Pérez FR. Trap crops and insectary plants in the order Brassicales. Annals of the Entomological Society of America. 2019;112:318-329. https://doi.org/10.1093/aesa/say043

[39] Santolamazza-Carbone S, Sotelo T, Velasco P, Cartea ME. Antibiotic properties of the glucosinolates of *Brassica oleracea var. acephala* similarly affect generalist and specialist larvae of two lepidopteran pests. Journal of Pest Science. 2016;89:195-206. DOI 10.1007/s10340-015-0658-y

[40] Giamoustaris A, Mithen R. The effect of flower colour and glucosinolates on the interaction between oilseed rape and pollen beetles. Entomologia Experimentalis et Applicata. 1996; 80:206-208. https://doi.org/10.1111/j.1570-7458.1996.tb00919.x|

[41] Ulmer BJ, Dosdall LM. Glucosinolate profile and oviposition behavior in relation to the susceptibilities of Brassicaceae to the cabbage seedpod weevil. Entomologia Experimentalis et Applicata. 2006;121:203-213. DOI: 10.1111/j.1570-7458.2006.00480.x

[42] Badenes-Pérez FR, Gershenzon J, Heckel DG. Plant glucosinolate content increases susceptibility to diamondback moth (Lepidoptera: Plutellidae) regardless of its diet. Journal of Pest Science. 2020;93:491-506. https://doi.org/10.1007/s10340-019-01139-z

[43] Stoner KA. Glossy leaf wax and plant resistance to insects in *Brassica oleracea* under natural infestation. Environmental Entomology. 1990;19:730-739. https://doi.org/10.1093/ee/19.3.730

[44] Lv H, Fang Z, Yang L, Zhang Y, Wang Y. An update on the arsenal: mining resistance genes for disease management of Brassica crops in the genomic era. Horticultural Research. 2020;7:1-18. DOI: 10.1038/s41438-020-0257-9

[45] Lema M, Cartea ME, Francisco M, Velasco P, Soengas P. Screening for resistance to black rot in a Spanish collection of *Brassica rapa*. Plant Breeding. 2015;131:607-613. DOI:10.1111/pbr.12293

[46] Iglesias-Bernabé L, Madloo P, Rodríguez VM, Francisco M, Soengas P. Dissecting quantitative resistance to *Xanthomonas campestris* pv. *campestris* in leaves of *Brassica oleracea* by QTL analysis. Scientific Reports. 2019;9:2015. doi.org/10.1038/s41598-019-38527-5

[47] Sotelo T, Lema M, Soengas F P, Cartea ME, Velasco P. In vitro activity of glucosinolates and their degradation products against Brassica-pathogenic bacteria and fungi. Applied and Environmental Microbiology. 2015;81:432-440. DOI: 10.1128/AEM.03142-14.

[48] Poveda J, Eugui D, Velasco P. Natural control of plant pathogens through glucosinolates: an effective strategy against fungi and oomycetes. Phytochemistry Reviews. 2020;19:1045-1059. https://doi.org/10.1007/s11101-020-09699-0

Chapter 2

Rapeseed-Mustard Breeding in India: Scenario, Achievements and Research Needs

Subhash Chand, Om Prakash Patidar, Rajat Chaudhary, Ranjit Saroj, Kailash Chandra, Vijay Kamal Meena, Omkar M. Limbalkar, Manoj Kumar Patel, Priya P. Pardeshi and Prashant Vasisth

Abstract

Brassica spp., commonly known as rapeseed-mustard, plays a significant role in the Indian economy by providing edible oils, vegetables, condiments and animal feed. Globally, India holds second and third position in rapeseed-mustard area under cultivation and production, respectively. However, anthropogenically accelerated climate change thwarts yield potential of rapeseed-mustard by employing abiotic (drought, flood, temperature variation and salinity) and biotic (disease and insects) stresses. Various approaches such as molecular breeding, pre-breeding, –omics and biotechnological interventions have been used to develop varieties for improved yield and oil quality, climate resilient and resistance or tolerance to abiotic and biotic stresses. In this context, this chapter highlighted the different cytoplasmic male sterility (CMS) sources and their potential use for hybrid development. At the end, this chapter also enlisted salient achievement by the government and non-government institutes and briefly described the future perspective for improvement of rapeseed-mustard in India.

Keywords: rapeseed-mustard, hybrid breeding, oil quality, pre-breeding, biotic and abiotic stress

1. Introduction

Brassica spp., commonly known as rapeseed-mustard, plays an important role in the Indian economy by providing edible oils, vegetables, condiments and animal feed [1]. Nine oilseeds are the primary sources of vegetable oil in India. Among them soybean (39%), groundnut (26%) and rapeseed-mustard (24%) contribute more than 88% of total oilseeds production in the country. However, rapeseed-mustard (31%) contributes maximum in terms of edible oil production followed by soybean (26%) and groundnut (25%) in the country [2].

Rapeseed-mustard is the third major edible oilseed crop of the world after soybean and palm oil. Globally, as per USDA during 2018-2019, it was grown over 36.6 million hectares and produced 72.4 MT with a productivity of 19.8 q/ha. Globally, India accounts 19.8% of total acreage and 9.8% of total production.

Species	Common name	Type of Pollination	Chromosome No. (2n)	Genome	Genome size (Mb)
B. juncea (L.) Czern.	Indian mustard	Often-self	36	AABB	~922
B. carinata A. Braun	Karan rai or Ethiopian mustard	Often-self	34	BBCC	—
B. napus L.	Gobhi sarson	Self and cross	38	AACC	~1130
B. nigra (L.) Koch	Black mustard	Cross	16	BB	~558
B. oleracea L.	Cabbage, cauliflower etc.	Cross	18	CC	~630
B. rapa L.	var. *brown sarson*	Lotni type: Cross Tora type: Self	20	AA	~485
	var. *toria*	Cross			
	var. *yellow sarson*	Self			
Eruca sativa	Taramira	Self	22	EE	—
B. alba Rab. (Syn. *Sinapis alba*)	White mustard	Self	24	SS	—

Table 1.
List of limited and importantly cultivated species of Brassica *species.*

Rapeseed-mustard (8.3 MT) is the third most important annual oilseed crop in India, next to soybean (13.6 MT) and groundnut (9.1 MT) [2]. In India, rapeseed-mustard is widely grown in diverse agro-climatic environments from North-East, North-West, Central to Southern states under different conditions such as sole crop/mixed crop, early/timely/late, rainfed/irrigated and saline or alkaline soils [3]. Based on average of 2014-2015 to 2018-2019 area and production data, major rapeseed-mustard growing states are Rajasthan (producing 44.9% of total rapeseed-mustard from 40.7% area), Madhya Pradesh (producing 11.3% from 11.9% area) and Uttar Pradesh (producing 10.6% from 11.2% area). Rapeseed-mustard crops in India comprise eight species *viz*., Indian mustard, toria, black mustard, yellow sarson, brown sarson, gobhi sarson, karan rai and taramira (**Table 1**).

2. Origin

Historically, the cultivation of *Brassica* spp. has been quoted in numerous ancient scriptures and believed to be cultivated on or prior to 5000 BC. It has also been reported that mustard crop had cultivated in Channhu-daro of Harrapan ancient civilization during 2300-1750 BC [4]. There is ambiguity in the history as the origin of *B. juncea* is concerned. It had been believed that center of origin for *B. juncea* is Middle-East, where putative parents *i.e. B. nigra* and *B. rapa* would have crossed with each other. Later on, it had been disseminated to other parts of the world such as Europe, Asia, and Africa etc. [5]. Today, there are two centers of diversity *i.e.* China and Eastern India based on the prevalence of their wild progenitors and relatives. At present, it has been proved that there are two geographical races *i.e.* Chinese and Indian of *B. juncea* based on molecular and biochemical studies [6].

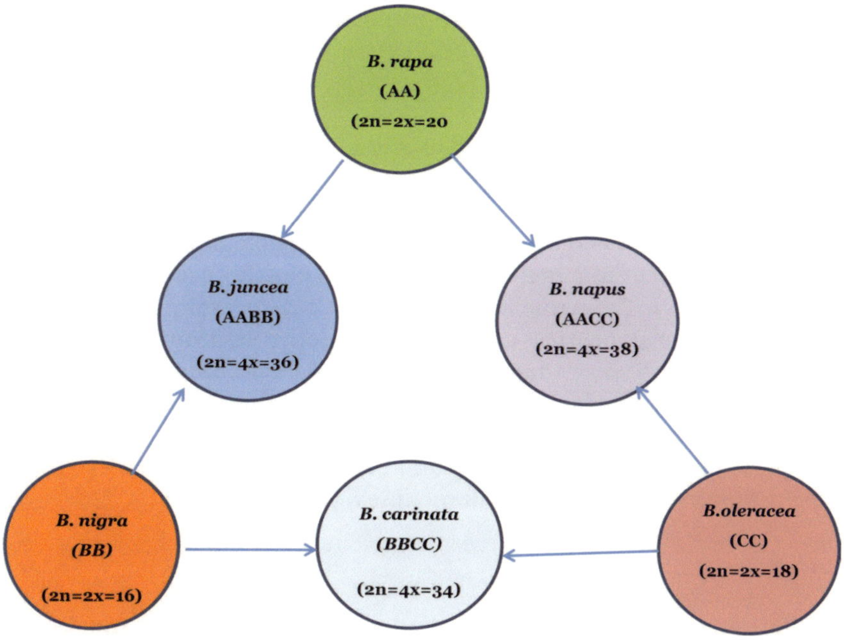

Figure 1.
U's triangle showing genetic relationship among six Brassica *species [7].*

In 1935, Nagaharu U [7] proposed a theory known as U's triangle to show genetic relationships based on artificial inter-specific hybridization experiments among six species, namely; *B. rapa, B. nigra, B. oleracea, B. carinata, B. napus* and *B. juncea*. As per theory, three allotetrapolyploid species (*B. napus, B. juncea and B. carinata*) were derived by natural hybridization of three basic diploid species (*B. rapa, B. nigra* and *B. oleracea*) followed by genome doubling (**Figure 1**). Nowadays, with the accomplishments of genome sequencing of *Brassica* taxa, this hypothesis has been increasingly accepted. Furthermore, it has been scientifically proved that allotetraploid *B. napus* and *B. juncea* had been derived from their diploid parents based on comparative genomic analysis and the results were in accordance with 'U' triangle [8].

3. Distribution

Brassicas include large number of crops under cultivation. Among them, the Indian mustard occupies maximum area (> 90%) and predominantly cultivated in North-Western states followed by some nontraditional areas of Central and Southern states of the country [1]. The lotni (cross-pollinated) and tora (self-pollinated) are two different ecotypes of brown sarson. Earlier one is mainly cultivated in temperate regions of the country such as parts of Jammu, Kashmir and hilly areas of Himachal Pradesh, whereas later one is cultivated in parts of Eastern Uttar Pradesh [3]. However, yellow sarson is predominantly cultivated in parts of Bihar, West Bengal and Orissa. Toria is mainly used as short period crop in parts of Bihar, West Bengal, Orissa and Assam. Whereas, it is grown as a catch crop in Haryana, Himachal Pradesh, Madhya Pradesh, Punjab, Uttarakhand and Western Uttar Pradesh. Taramira, relatively more drought tolerant, is cultivated in drier parts of Rajasthan, Uttar Pradesh and Haryana. However, karan rai and gobhi sarson have limited area under cultivation in India [1].

4. Breeding approaches in rapeseed-mustard

4.1 Abiotic stresses

Plant stress factors can be elucidated as any adverse condition or substance that affects the growth, reproduction, metabolism and development of the plant [3]. Acclimatization or hardening refers to exposure of unfavorable environmental circumstance to the plant and thereby results into physiological adjustment that protects it from injury or impaired growth which is mostly occurred due to environmental stresses [9]. There might be fixed genetic changes if plant faces several generations under constant stress condition by selective environmental pressure and thereby population show adaptation to changed environment. Abiotic factors are the main yield-limiting factors for crop plants including rapeseed-mustard. The major abiotic factors are- moisture variation (drought and flood), temperature variation (heat, cold and frost), salinity and heavy metal that adversely affect the metabolic pathways and thereby result into yield penalty.

4.1.1 Drought stress

Globally, rapid climate change under anthropogenic accelerated interventions crafts drought a major menace to the agricultural production system and consequently has a great challenge to the global food and nutritional security. Plants have different ways to synergies with drought stress such as modifications in plant growth, behavior, morphology, and physiology. In *Brassica*, drought tolerance is a complex trait and thereby associated with different traits; and can be evaluated by various indicators. Moreover, it is difficult to choose all the exiting indicators at a time to use in breeding programs for crop improvement. Drought can adversely affect plant growth at various stages from seed germination to reproduction and flowering to harvesting, and ultimately results into oil and yield penalty [3]. Prolonged drought reduces chlorophyll content mostly due to impaired functioning of thylakoid membrane and heavy loss of pigments [10]. In the context, the pattern of gene expression of those traits which are associated with osmotic balance, water transport, damage repair and oxidative stress will be altered by prolonged drought stress (**Table 2**). Thus, drought is one of the major factors to reduce potential yield of crop plants and introgression of traits from wild relatives can be used for the development of drought resilient cultivars in rapeseed-mustard.

4.1.2 Salt stress

Recent advances in molecular breeding have been characterized and genetically mapped various salt related genes in plants. Gradual increase of the understanding of several biochemical, and physiological mechanisms and pathways of salt related genes has made it easy to develop genetically improved varieties which are more resilient and high yielding under salinity stress. In this context, transgenic approaches have also been used to know the effect of salt tolerant genes into the different genetic background by up-regulating or down-regulating genes under salt stress [33]. The progress under salt tolerance is great in major agricultural crops such as wheat, rice, mustard and tomato. A large number of gene (s)/*QTLs* have been mapped as well as cloned [33]. As *Brassica* crops are concerned, there are limited studies on salt regulating genes or *QTLs* across the world. In India, only limited salt tolerant varieties have been developed so far such as “CS56” and breeding approaches are not as much successful as to other stresses [3]. It is need of the hour to understand the mechanism of salt tolerance and to identify stable salt tolerance genotypes from available genetic resources

Species	Gene/s	Function	Tolerance	References
Arabidopsis	*DREB1A*	Dehydration response element binding protein	Drought, salt and freezing	[11]
	SOS1	Plasma membrane-bound Na+/H+ antiports	Salt	[12]
	AtNHX1	Vacuolar Na+/H+ antiporter	Salt	[13]
	AtHKT1	Na + transporter	Salt	[14]
	FTA	Farnesyltransferase	Drought	[15]
	AtFTB	β-subunit of Farnesyltransferase	Drought	[16]
Arthrobacter globiformis	*codA*	Choline oxidase	Salt	[17]
B. rapa	*BrERF4*	Ethylene-responsive factors	Drought and salt	[18]
	BrGI	Reduced expression of GI, enhanced salt tolerance	Salt	[19]
B. napus	*AtDWF4*	Enhanced defense gene expression	Drought and heat	[20]
	*BnNHX1*and *BnHKT*	Salt-responsive genes	Salt	[21]
	BnLEA4-1	Late-embryogenesis abundant proteins in group 4	Salt	[22]
	BnLAS	Transcriptional regulator members in GRAS family	Drought	[23]
	DREB	Improving the abiotic stress tolerance	Salt	[24]
	BnSIP1-1	Played roles in ABA synthesis and signaling	Salt and Osmotic	[25]
	AnnBn1	Membrane-binding proteins for Ca2+	Drought	[26]
B. oleracea var. *botrytis*	*APX, SOD*	Protect from oxidative stress	Salt	[27]
B. juncea cv. varuna	*Glyoxalase I* *Lectin*	Catalyze the detoxification of a highly cytotoxic metabolite methylglyoxal to d-lactate	Drought and salt	[28]
B. juncea	*BrECS*	Glutamylcysteine synthetase	Salt	[29]
	AtLEA4-1	AtLEA4-1 LEA4 protein	Salt	[30]
	Gly I	Detoxification of methylglyoxal	Salt	[31]
	AnnBj2	Upregulated expression of ABA-dependent (*RAB18*) and ABA independent (*DREB2B*) genes	Salt	[32]

Table 2.
Brief summary of abiotic stress tolerance associated genes and their functions.

by extensive screening methods to use them in breeding programs. Researchers have done excellent work on ion homeostasis and osmolytes regulation by using transgenic approach in *Brassica* crops [34] and identified few candidate genes (**Table 2**).

Apparently, both drought and salinity stress have few similarities in plants. Both stresses are primarily responsible for cellular dehydration, which removes water from the cytoplasm into the intercellular space [35]. Based on the functional similarity of both the stresses in plants, it can be concluded that plants have almost identical mechanism to deal with both stresses. In the present scenario, researchers are extensively working on model plant *i.e. A. thaliana* to understand the genetics of salt and drought stress tolerance, which can positively help to develop tolerance cultivars in *Brassica* spp. and will improve agronomically important traits [36].

4.1.3 Heat stress

As the global warming is increasing due to unwarranted human activities, heat stress has become a major factor to hamper plant growth and development in agricultural crops including rapeseed-mustard. Early sowing of Indian mustard, have various advantages as enlisted by Kaur and coworkers [37] but high temperature during the germination stage leads to reduction in the plant emergence and poor plant stand. The yield potential of Indian mustard was significantly reduced under late sown condition compared to timely sown due to terminal heat stress [38]. The reduction in emergence of Indian mustard due to hot soils can lead to substantial economic losses [39]. Where irrigation is available and multiple cropping system followed, especially in Central and North-Western plain zones, sowing of the mustard crop is delayed up to end of November due to late vacation of *Kharif* crop, leads to exposure of the crop to high temperature at maturity.

Rapeseed-mustard is adversely affected by heat stress (35/15 °C) at the early stage of flowering. Moreover, yield penalty can be avoided if high temperature occurs during early pod formation. In this context, *B. rapa* is more sensitive to high temperature whereas *B. juncea* and *B. napus* are equally affected [40]. It has been reported that optimal temperature for *B. napus* is lower than *B. juncea* and *B. rapa* [41]. Generally, as temperature increased, the number of pods produced by the plants increased and seed weight decreased. High temperature has a direct effect on the formation of reproductive organs. More research is needed under controlled environments to identify the critical temperature, sensitive reproductive organ stage, source-sink relationship, and genotypic variations for heat stress tolerance and must be verified under natural conditions [42].

4.1.4 Low temperature stress

Freezing injury has adverse effect on plant growth and development, and thereby leads to yield penalty. Seed germination is seriously affected by low temperature. Plant stress hormones such as Brassinolide (BR) regulate plant physiological pathways and helps in plant protection to combat low temperature stress [43]. Exogenous application of BR increased cold stress tolerance in *A. thaliana* and *B. napus* [44]. In this context, BR increases chlorophyll content, PS-II, antioxidant enzymatic activities and protect photosynthetic membrane system from oxidative damage [45]. It has been reported that accumulation of reactive oxygen species such as superoxide anion, hydrogen peroxide, singlet oxygen and hydroxyl radical is high under cold stress, and thereby causes oxidative stress in plants which leads to cell death [46]. The *B. rapa* has been reported more cold tolerance than *B. napus*. The impact of heat stress is high than cold stress because of inactivation of RuBisCO and/or other associated enzymes under heat stress. Intriguingly, *B. oleracea* is cold tolerant due to its acclimatization in cold regions of Europe, where summer temperature is also low and crop had domesticated since long back.

Thus, acclimatization, domestication, adaptive trans-generational plasticity and genetic adaptation phenomenon can work simultaneously to abiotic stress tolerance in *Brassica* species.

4.2 Biotic stresses

A number of biotic stresses adversely affect the yield potential of rapeseed-mustard in India. The major diseases are- Alternaria blight (*Alternaria brassicae* and *A. brassicicola*), white rust (*Albugo candida*), stem rot (*Sclerotinia sclerotiorum*), Rhizoctonia rot and downy mildew (*Peronospora brassicae*); and major insect pests are- aphid (*Lipaphis erysimi*), mustard saw fly (*Athalia proxima*) and painted bug (*Bagrada hilaris*). There are several methods to control insect and disease incidence such as application of pesticides, fungicides, biological agents and other non-chemical techniques. However, the most economic, eco-friendly and cheap way to mitigate these menaces are to use of resistant or tolerant cultivars through convention and molecular breeding approaches.

4.2.1 Alternaria blight

The yield potential of *Brassica* spp. is adversely affected by Alternaria blight [*Alternaria brassicae* (Berk) Sacc.] disease. The pathogen can affect the host plant at all stages of growth and highest disease severity was observed during rainy season. The *B. juncea* and *B. rapa* are more susceptible than *B. carinata* and *B. napus* to Alternaria blight. The researchers have reported several sources of disease tolerance such as *B. juncea* cv. Divya, and wild species such as *Sinapis alba* L., *B. maurorum*, *Diplotaxis berthautii* and *D. erucoides* etc. [47]. Higher concentration of phenolic compounds (polyphenol peroxidase, oxidase and catalase), low N content, higher leaf sugar content, and more leaf wax deposition have been reported to deliver resistance to plants against Alternaria blight disease [48]. Pre and post fertilization barriers are major concern while using wild relatives and progenitors as donor source in rapeseed-mustard breeding programs. However, limited sources of *B. juncea* (PHR 2, RC781, Divya, PAB 9534, and EC 399301) have been reported tolerance against this disease and extensively being used in breeding programs [3].

4.2.2 White rust

White rust [*Albugo candida* (Pers.) Kuntze] is a destructive disease in *B. juncea* and *B. rapa*; and significantly reduces potential yield up to 60% in mustard [49]. Forty-nine races of *A. candida* have been reported in India based on their infectivity on different *Brassica* spp. and their cultivars [50]. Most of the varieties under Indian mustard are susceptible to white rust whereas *B. carinata* and *B. napus* demonstrate high degree of resistance. Thus, gene introgression from *B. carinata* and *B. napus* to *B. juncea* through interspecific hybridization is essential for development of resistant or tolerant cultivars in the country [51]. The varieties bred for disease tolerance are- JM-1, JM-2, DMH-1 and Basanti etc.

4.2.3 Sclerotinia rot

In rapeseed-mustard, Sclerotinia rot disease is triggered by *Sclerotinia sclerotiorum* and adversely affects plant growth and development. The disease has turned form minor significance to major one since last decade due to change in climatic condition. Pre-mature ripening is the cause of the disease. The pathogen has an array of alternate host therefore breeding for disease resistant is difficult [3].

4.2.4 Insect (Aphid)

Mustard aphid (*Lipaphis erysimi*) is one of the major insect pests in rapeseed-mustard and adversely affects plant growth, development, and reproduction; and thereby results into yield penalty. They are also act as vector for plant viral diseases such as turnip mosaic virus. There are several methods to identify resistant source for aphid resistance/tolerance in *Brassica* family such as based on seedling survival, aphid fecundity, and aphid infestation index etc. Some genotypes of *B. juncea* such as Glossy B-85, RH 7847, and T 6343 were reported more tolerant to aphid infestation. *B. campestris* is more susceptible to aphid infestation than *B. juncea* and *B. carinata* [3].

4.3 Oil quality improvement

The oil quality for human consumption is determined by its fatty acid composition and concentration. Seed oil with high proportion of unsaturated fatty acid, particularly 16 and 18 carbon chain, is considered suitable for human consumption as edible oil. Rapeseed-mustard is mostly used as oilseed crop in India and its seed contain 35-45% oil content with 92-98% triacylglycerol of fatty acids (C16-C22). Seed oil contains lowermost saturated fat and possesses high proportion of essential fatty acid such as linoleic (C18:2) and linolenic (C18:3) which are not synthesized by human body. Linolenic acid is an essential dietary fatty acid; however, its higher concentration reduces shelf-life of oil because of auto-oxidation [3]. Erucic acid (C22:1) comprises almost 50% of total seed oil fatty acid in rapeseed-mustard and is undesirable for human consumption due to its adverse role in myocardial conductance and increase the level of blood cholesterol. The level of detrimental saturated fatty acid is less in rapeseed-mustard compared to other edible oilseed crops. The major constrains in seed oil are- erucic acid and glucosinolates [52]. Therefore, reduced concentration of glucosinolates and erucic acids is one of the important objectives in quality amelioration of Indian mustard seed oil. It has been reported that genetic inheritance of glucosinolates is complex and mostly are aliphatic (methionine derived) in nature in *B. juncea*. Genetic control of total glucosinolates in *B. juncea* has been reported to be under two major genes [53], multiple additive alleles at a single locus with maternal effects involved [54], six to seven genes [55] and up to five major QTLs [56] based on molecular mapping information.

The rapeseed-mustard varieties with low erucic (<2%) and glucosinolates (<30 μ mole/g of defatted cake) are termed as double zero ("00"). The term single zero ("0") is used when variety contains only one factor either low erucic (<2%) or glucosinolates (<30 μ mole/g of defatted cake). In this context, several efforts have been made to improve oil quality of rapeseed-mustard in India since last three decades. In India, first low erucic acid ("0") variety was LES-39 (Pusa Karishma) followed by LES-1-27 (Pusa Mustard 21), LET-18 (PM 24), and LET-17 (PM-22) in *B. juncea,* whereas double zero variety was Pusa Double Zero Mustard 31 (PDZM-1).

4.4 Hybrid breeding

Rapeseed-mustard exploits high level of heterosis but employ difficulty in seed production due to complex flower structure, presence of self-compatibility and thereby self-pollination in nature, however crop also enjoyed cross-pollination (30%) by pollinators such as honey bees. The extent of heterosis was reported by Sun [57] in rapeseed-mustard during early forties and was pioneer to begin with hybridization for exploitation of hybrid vigor. Subsequently, Ogura

[58] had successfully transferred male sterile cytoplasm from radish (*Raphanus sativus* L.) to *B. juncea*. In this context, several cytoplasmic male sterility systems have been reported such as *tour* [59] in *B. napus*, *oxyrrhina* [59], *siifolia* [60], *trachystoma* [61], *moricandia* [62], *catholica* [63], *alba* [62], *lyratus* [64], *canariense* [65], *erucoides* [66], *126-1* [67] and *barthauti* [68]. Transgenic male sterility (barnase-barstar system) system was also used for exploitation of heterosis and development of hybrid varieties [69, 70]. It has been reported that large number of sterile cytoplasm is available, however only few can be utilized in heterosis due to lack of adequate and efficient fertility restoration system. Therefore, ICAR sponsored project (1989) "Promotion of Research and Development Efforts on Hybrids in Crops" which aimed for systematic and coordinated efforts for hybrid development in rapeseed-mustard in India with two CMS systems (*ogu* and *tour)* in *B. juncea* while *polima* in *B. napus*.

In India, heterosis was first reported in brown sarson (*B. rapa*) by Singh and Mehta [71]. It has been reported that the extent of heterosis is 13 to 99% in *B. juncea*, 10 to 72% in *B. napus*, 25 to 110% in *B. rapa*. Generally, hybridization between genetically distinct groups exploits high level heterosis than within group. Exploitation of high level of heterosis in plants necessitates large and usable heterosis, effective pollination control mechanism, and profitability of seed production [70]. Thus, there is urgent need to improve genetic gain and heterosis in rapeseed-mustard; genetic variability, in terms of variety, can be tested for 2-3 years across the centers in the country through All India Coordinated Research Project [72] and by result of high yielding, stress tolerance and stable variety would be produced.

4.4.1 Cytoplasmic male sterility and hybrids

A large number of CMS systems are available in rapeseed-mustard such as *Raphanus/ogu*, *tour*, *oxyrrhina*, *siifolia*, *trachystoma*, *moricandia*, *catholica*, *lyratus*, *canariense*, *erucoides*, and *barthauti* (**Table 3**). All the CMS sources cannot be directly used in hybridization programme due to their negative effects on plant growth and development such as chlorosis (*ogura*, *oxyrrhina* and *moricandia*), impaired flower opening (*tour*, *trachystoma* and *lyratus*), and also absence of fertility restoration. The chlorosis of three systems (*ogu*, *oxyrrhina*, *moricandia*) had been cured through somatic hybridization by fusing protoplast of chlorotic sterile and normal green plant [74]. The fertility restorer genes (*Rfs*) were identified in five CMS systems *viz*. *trachystoma*, *moricandia*, *catholica*, *canariense* and *lyratus* in their respective cytoplasmic donor species and restorer can be isolated simultaneously during transfer of sterile cytoplasm.

The success of hybridization programme, by using CMS system, depends upon availability of efficient fertility restoration. In rapeseed-mustard, the utmost used CMS system in India are-*Raphanus/ogu* CMS system, *B. tournefortii* CMS system, *Moricandia arvensis* CMS system, and *Erucastrum canariense* CMS system. In India, the first commercial hybrid PGSH 51 (*B. napus*) was released in 1994 based on *tour* CMS and yield was increased by 18% over the best hybrid check. The other hybrids are as follow- Hyola 401 hybrid (2000) was based on *pol* CMS system, NRCHB-506 (2008) on *mori* cytoplasm, DMH-1 (2008) on *126-1* CMS, and PAC-432 (2009) on *ogu* cytoplasm etc. The genetic engineering techniques had also utilized for the development of male sterile system to exploit the heterosis in rapeseed-mustard and develop the barnase-barstar male sterile system [69, 70]. Hybrid DMH-11 was developed by Delhi University in India which became India's first transgenic hybrid through barnase-barstar system. But DMH-11 was not released for commercial cultivation due to resistance from environmental activist in thought of its harm to environment.

CMS system	Discovered by	Year	Fertility restoration
Raphanus/ogu	Ogura [58]	1968	Restorer gene is available in *B. juncea*
tour	Rawat and Anand [59]	1979	Available in *B. napus*
oxyrrhina	Prakash and Chopra [73]	1988	No restoration available
siifolia	Rao and coworkers [60]	1994	No restoration available
trachystoma	Kirti and coworkers [61]	1995	Single dominant gene available for restoration
moricandia	Prakash and coworkers [62]	1995	Single dominant gene reported for restoration
catholica	Kirti and coworkers [63]	1995	Reported but not in use
alba	Prakash and coworkers [62]	1995	Available in *B. napus*
lyratus	Banga and Banga [64]	1997	Reported but not in use
canariense	Prakash and coworkers [65]	2001	Reported but not in use
erucoides	Bhat and coworkers [66]	2006	Reported but not in use
126-1	Sodhi and coworkers [67]	2006	Reported in *B. napus*
barthauti	Bhat and coworkers [68]	2008	Reported but not in use

Table 3.
Important sources of CMS in rapeseed-mustard for hybrid seed production.

4.5 Pre-breeding

Wild progenitors and wild relatives are to be known as repository of valuable traits (quality, agronomic, biotic and abiotic stress tolerance) in crop plants but cannot be introgressed into the cultivated ones due to linkage drag, and cross-incompatibility barriers. Pre-breeding helps to identify the useful traits in wild germplasm and employ its use in breeding programs. The major objective of pre-breeding is to introduce new variation into the species of interest with minimum linkage drag. Molecular markers would play a great role to accelerate the breeding cycle, reduction in cost and time, and increase in the efficiency of introgression in pre-breeding programs [75].

Globally, India (15%) ranked second after China (17%) in terms of repository of *Brassica* germplasm. In India, National Bureau of Plant Genetic Resources (NBPGR) has contributed 4095 indigenous and 3401 exotic rapeseed-mustard accessions from 1986-2006 [76]. All the efforts have resulted into the collection of a total of 14,722 accessions of cultivated, wild relatives, wild progenitors and related species [3]. There is a wide gap between available germplasm in gene banks and its utilization in the breeding programs due to lack of available identified traits. Thus, there is urgent need to broaden the plant genetic diversity to combat anthropogenically accelerated climate change in the near future.

5. Biotechnological approaches

Rapeseed (*B. napus*), cultivated in temperate climate, have been believed to originate by natural hybridization between *B. oleracea* and *B. rapa*. *B. napus* was resynthesized by protoplast fusion of *B. oleracea* and *B. rapa* to widen genetic diversity and alter oil content. The biotechnical intervention was used either to increase of genetic variability or transfer of desirable traits from other related species such wild relatives, wild progenitors or other unrelated crops to improve yield potential of crop which were not possible due to conventional or classical breeding methods.

5.1 Anther culture

Pollen culture can be used to develop stable homozygous lines by double haploid (DH) technique to improve agronomic traits in *B. juncea*. Improvement in culture condition and associated factors, which are limiting factor for embryo production, tend to increase efficiency of microspore culture or anther culture in *B. juncea* [77]. It has been reported that microspore culture is more successful than anther culture due to better response of genotypes for embryo culture. Microspore culture can be used for gene transfer, biochemical studies, and modification of fatty acid profile through mutagenesis [77]. The major factors which affect doubled haploid production are- isolation of microspore, culture media, embryo selection, plant regeneration, and chromosomal duplication. In India, there is no variety under cultivation of this technique.

5.2 Somaclonal variation

Somaclonal variation can be defined as genetic variation in somatic cells due to chromosomal rearrangement and regeneration of variable plants from callus by plant tissue culture. Furthermore, *B. juncea* variety Prakash produced multiple shoots in cotyledonary callus when high cytokinin and low IAA concentration was used in MS media [78]. A large genetic variation has been created in *B. juncea* by tissue culture through induced somaclonal, chemical mutagens, and gamma rays induced variation. For example, somaclone- SC-122 was developed with improvement of five traits which were associated with yield improvement [79]. In India, Pusa Jai Kisan (Bio-902) was first somaclonal derived variety in 1993 by using Varuna as a parent and yield was improved by 17.4% over the parent.

5.3 Protoplast culture

Protoplast, cell without cell wall, culture induces protoclonal variation and creates stable genetic variability in rapeseed-mustard by using tissue culture technique. This technique was used *B. juncea* cv. RLM-198 by using V-47 media for production of somatic embryo and organogenesis. This method can be used for those *Brassica* species where hybridization is not possible and will help to create genetic variability for betterment of crop improvement.

5.4 Transgenic plants

In crop species, transgenic plants have been developed by using the recombinant DNA technology. It has been widely used to transfer alien gene/chromosomal segment to the recipient parent where naturally gene of interest is absent for betterment of mankind. Various direct and indirect methods have been used for gene transfer in crop plants including rapeseed-mustard and mostly used direct method is *Agrobacterium* mediated gene transfer for seed yield, seed quality, biotic and abiotic stress tolerance and desirable agronomic traits [80]. As earlier mentioned, transgenic male sterility system was used for production of hybrids in India. Thus, these biotechnological interventions can solve the problems of conventional breeding which are mainly associated with hybridization and selection.

5.5 -Omics approaches

The world of –omics is vast and covers several disciplines such as genomics (total DNA content of organism), transcriptomics (deals with total RNA content), proteomics (deals with total proteins), and metabolomics (total metabolites of an individual). Being amphidiploid and tetraploid in nature, both *B. juncea* and *B. napus* need -omics approaches to understand the trait based genetics for improvement of these crops.

5.6 Genomics

Linkage mapping and association studies were used to identify the genomic locations of a particular trait of interest. Genomic locations were identified based on molecular markers in *Brassica* spp. For example, Mukherjee and coworkers [81] mapped genes governing white rust resistance using BSA in *B. juncea*. Padmaja and coworkers [82] mapped seed coat color gene and identified microsatellite markers, *Ra2-A11*, *Na10-A08* and *Ni4-F11* linked to seed coat color in *B. juncea*. Furthermore, Liu and coworkers [83] dissected genetic architecture for glucosinolates accumulation in seed and leaves using GWAS in *B. napus*. Kaur and coworkers [84] carried out genome wide association mapping and candidate gene analysis for pod shatter resistance in *B.juncea*. Comparative mapping was also used in rapeseed-mustard for different agronomic and quality traits. For example, Cai and coworkers [85] identified candidate gene- *BnAP2* for seed weight in *B. napus* by using comparative mapping with *A. thaliana*. Bisht and coworkers [86] identified candidate genes, *BjuA.GSL-ELONG.a*, *BjuA.GSL-ELONG.c*, *BjuA.GSL-ELONG.d*, *BjuA.GSL-ALK.a* and *BjuA.Myb28.a* for glucosinolates biosynthesis through comparative mapping among *A. thaliana*, *B. oleracea* and *B. juncea*. Genomics has been extensively used for evolutionary studies in *Brassica* spp. Couvreur and coworkers [87] used *nad4 intron 1* marker for phylogenetic analysis to study temporal diversification and establishment of evolutionary pattern in the mustard family. Furthermore, Augustine and coworkers [88] isolated four *BjuCYB83A1* genes from *B. juncea*, which involved in glucosinolates synthesis and through phylogenetic and divergence analysis they have revealed that these genes have evolved via duplication and hybridization of two diploid *Brassica genomes i.e. B. rapa* and *B. nigra*.

5.7 Transcriptomics

Transcriptomics contributes the comprehensive understanding about the gene expression, through which it is easy to allocate gene function and its effect on any organism. It has been used for expression studies, gene silencing, and genome editing in *Brassica* spp. for example, Heng and coworkers [89] identified *orf288* gene associated with male sterility in *B. juncea* through expression analysis of *orf288* transcript. Bhattacharya and coworkers [90] studied down regulation of BjAGPase and seed specific expression of *AtWRI1* gene of *Arabidopsis* in order to increase seed lipid content in *B. juncea*. Savadi and coworkers [91] increased seed weight and seed oil content in Indian mustard through seed specific overexpression of *DGAT1* gene of *A. thaliana.* Zhao and coworkers [92] carried out RNAi mediated gene silencing of *mutS homolog1* which results in male sterility in *B. juncea* due to sub-stoichiometric shifting in *ORF220*. Zheng and coworkers [93] carried out gene knockout experiment through CRISPR/Cas9 in *BnaMAX1* homologs of *B. napus*, which resulted in reduction in plant height and increase in branch number.

5.8 Proteomics

Proteins are the ultimate products which confer the gene function and govern the phenotypic expression to an individual. Proteomics approaches such as protein expression profiling and comparative proteomics analysis were used to study the gene function in *Brassica* spp. For example, Mihr and coworkers [94] used "*Tournefortii*" CMS system of *B. napus* to study protein content of mitochondrial compartments in male sterile and fertile NILs. Mohammadi and coworkers [95] performed comparative proteome analysis in rapeseed seedlings for root traits under draught stress and concluded that proteins such as H^+ ATPase, HSP 90 and EF2 play a key role in draught tolerance. Yousuf and coworkers [96] identified salt stress responsive proteins in the shoots of Indian mustard genotypes through comparative proteome analysis approach. Yousuf and coworkers [97] studied different protein expression profiles of N_2 efficient and N_2 inefficient Indian mustard in response to elevated CO_2 and low N_2.

5.9 Metabolomics

Recent efforts in metabolomics have been directed to improve quality and yield of any crop. An integration of metabolomics with other approaches establishes an important relevance in crop improvement. However, metabolomics has not exploited much in mustard breeding, so it would be an emerging field of research for *Brassica* improvement. Few studies have been carried out in *B. juncea*. For example, Sinha and coworkers [98] performed metabolic engineering of fatty acid biosynthesis in order to improve nutritional quality of seed oil in Indian mustard. Kortesniemi and coworkers [99] investigated seed metabolomics using NMR in *B. napus* and *B. rapa* and found that unsaturated fatty acids, sucrose and sinapine were most discriminating metabolites.

6. Achievements

In India, 189 rapeseed-mustard varieties (118 Indian mustard; 7 karan rai; 14 gobhi sarson; 24 toria; 15 yellow sarson; 3 brown sarson; 1 black mustard; 7 taramira) were developed and released and some of them are enlisted in **Table 4**. Several CMS based hybrids were developed by government and non-government institutes. A total of 7029 accessions comprising toria (508), Indian mustard (4,600), yellow sarson (548), gobhi sarson (146), brown sarson (108), karan rai (232), taramira (67), *B. caudatus* (04), *R. caudates* (01), *B. rugose* (30), *B. nigra* (22), *S. alba* (01), *Crambe* spp. (02), and *Lapidium* spp. (02) were maintained through appropriate mating system at various coordinated centers in the country [100]. As seed oil quality is concerned, low glucosinolates content was transferred from agronomically poor exotic genetic stock of *B. juncea*, BJ-1058 to the genetic background of high yielding mustard varieties. Genetics of fatty acid profile and glucosinolates content has been worked out and gene pool for high oil content and disease resistance were developed.

7. Future outlook and strategy

To fulfill the demand of edible oil for ever increasing population, constant efforts are needed for higher production and productivity by conventional, molecular or biotechnological approaches in the country. Genetic variability is the prerequisite for crop improvement program. Moreover, there is imperative need to

Stress/situation/condition	Recommended varieties
Salinity	Indian mustard: CS-54, Pusa Vijay, NRCDR 2, CS 234-4, CS-52, Narendra Rai-1, NRCDR 601
High temperature	Indian mustard: Urvashi, RGN 13, Pusa Agrani, Kanti, PM 26, PM 27, DRMR 1165-40, NRCDR 2, NRCDR 601
High oil content	Indian mustard: Narendra Swarna Rai 8
Earliness	Indian mustard: Kanti, Narendra Ageti Rai 4, Pusa Agrani, Pusa Mahak, DRMR 150-35; Yellow sarson: NRCYS 05-01
Intercropping	Indian mustard: RH-30, RH781, Vardan
Non-traditional areas	Indian mustard: Pusa Agrani, Pusa Jai kisan, Gujarat Mustard 2, Pusa Mahak (for north-east only)
Late sown	Indian mustard: Ashirwad, RLM 619, SwaranJyoti, Vardan, Navgold, NRCHB 101
Frost tolerance	RGN13, RH-781, SwaranJyoti
Drought (Rainfed)	Indian mustard: RH-819, RH 781, GM1, Pusa Bahar, Pusa Bold, Aravali Mustard, Sej-2, JD-6, Geeta, RGN-48, RL-99-27, Shivani, PBR-9 Karan rai: Pusa Aditya, DRMR 150-35, Pusa Swarnim
Irrigated	Indian mustard: PM-28, DRMRIJ 31
Low erucic acid /glucosinolates	Indian mustard: Pusa Karishma, Pusa Mustard 21, PM 22 Gobhi Sarson: Hyola 401, GSC 5, GSC 6, NUDB 26-11, Teri Uttam Jawahar, PM 24
White rust	Indian mustard: Basanti, JM 1, JM 2, Maya, Pusa Jagannath
Powdery mildew and Alternaria blight	Indian mustard: DRMR 150-35, NRCDR 2, NRCDR 601
Wider adaptability	Indian mustard: Pusa Bold

Table 4.
Improved varieties of Indian mustard for specific environmental conditions.

diversify the genetic base of varieties by utilization of exotic germplasm as well as other wild and related species. In this context, combination of conventional plant breeding with biotechnological tools can be used for development of high yielding varieties with good oil quality and tolerance against biotic and abiotic stresses. Global warming and the climate change are very critical challenges in the near future. Efforts to develop climate resilient crop cultivars are the need of the hour. Marker assisted selection (MAS), functional genomics, phenomics, proteomics and metabolomics are the next step to develop varieties for drought and heat tolerance and breeding programs must be reoriented to meet the future challenges. Nowadays, omics breeding has emerged as a novel concept in crop improvement and upcoming era will be dominated by this approach as it is more robust and rapid as compared to conventional breeding.

Conflict of interest

"The authors declare no conflict of interest."

Author details

Subhash Chand[1*], Om Prakash Patidar[1], Rajat Chaudhary[1], Ranjit Saroj[1], Kailash Chandra[2], Vijay Kamal Meena[1], Omkar M. Limbalkar[1], Manoj Kumar Patel[1], Priya P. Pardeshi[1] and Prashant Vasisth[1]

1 Division of Genetics, ICAR- Indian Agricultural Research Institute, New Delhi, India

2 Department of Plant Breeding and Genetics, Sri Karan Narendra Agriculture University, Jaipur, Rajasthan, India

*Address all correspondence to: subhashchand5415@gmail.com

References

[1] Jat RS, Singh VV, Sharma P, Rai PK. Oilseed *Brassica* in India: Demand, supply, policy perspective and future potential. OCL. 2019; 26: 8

[2] MAFW. Annual report 2018-19 by Ministry of Agriculture and Farmers Welfare, Government of India. 2019; Retrieved from http://agricoop.nic.in/sites/default/files/AR_2018-19_Final_for_Print.pdf

[3] Chauhan JS, Singh KH, Singh VV, Kumar S. Hundred years of rapeseed-mustard breeding in India: accomplishments and future strategies. Indian J Agric Sci. 2011; 81 (12):1093-1109

[4] Allchin FR. In the domestication and exploitation of plants and animals eds; P.J. Ucko and GW Dimbledy, London. 1969. pp. 323-329

[5] Hemingway JS. Mustard: *Brassica* spp. and *Sinapis alba* (cruciferae). In Evolution of Crop Plants. Simmonds (Ed,), Longmans, London. 1979; pp. 56-59

[6] Song KM, Osborn TC, Williams PH. *Brassica* taxonomy based on nuclear restriction fragment length polymorphisms (RFLPs). Theoretical and Applied Genetics. 1988; 75(5): 784-794

[7] Nagaharu U. Genome analysis in *Brassica* with special reference to the experimental formation of *B. napus* and peculiar mode of fertilization. Jpn J Bot. 1935; 7(7): 389-452

[8] Yang J, Liu D, Wang X, Ji C, Cheng F, Liu B, Hu Z, Chen S, Pental D, Ju Y, Yao P. The genome sequence of allopolyploid *Brassica juncea* and analysis of differential homoeolog gene expression influencing selection. Nature Genetics. 2016; 48(10): 1225-1232

[9] Crisp PA, Ganguly D, Eichten SR, Borevitz JO, Pogson BJ.Reconsidering plant memory: Intersections between stress recovery, RNA turnover, and epigenetics. Science Advances. 2016; 2(2): e1501340

[10] Champolivier L, Merrien A. Effects of water stress applied at different growth stages to *Brassica napus* L. var. *oleifera*on yield, yield components and seed quality. European Journal of Agronomy. 1996; 5(3-4): 153-160

[11] Kasuga M, Liu Q, Miura S, Yamaguchi-Shinozaki K, Shinozaki K. Improving plant drought, salt, and freezing tolerance by gene transfer of a single stress-inducible transcription factor. Nat Biotechnol. 1999; 17(3): 287-291

[12] Shi H, Ishitani M, Kim C, Zhu JK. The *Arabidopsis thaliana* salt tolerance gene SOS1 encodes a putative Na^+/ H^+antiporter. Proceedings of the national academy of sciences. 2000;97 (12): 6896-6901

[13] Zhang HX, Hodson JN, Williams JP, Blumwald E. Engineering salt-tolerant *Brassica* plants: characterization of yield and seed oil quality in transgenic plants with increased vacuolar sodium accumulation. Proceedings of the National Academy of Sciences. 2001; 98(22): 12832-12836

[14] Berthomieu P, Conéjéro G, Nublat A, Brackenbury WJ, Lambert C, Savio C, Uozumi N, Oiki S, Yamada K, Cellier F, Gosti F. Functional analysis of AtHKT1 in *Arabidopsis* shows that Na+ recirculation by the phloem is crucial for salt tolerance. The EMBO Journal. 2003; 22(9): 2004-2014

[15] Wang Y, Ying J, Kuzma M, Chalifoux M, Sample A, McArthur C, Uchacz T, Sarvas C, Wan J, Dennis DT, McCourt P. Molecular tailoring of farnesylation for plant drought tolerance and yield protection. The Plant Journal. 2005; 43(3): 413-424

[16] Wang Y, Beaith M, Chalifoux M, Ying J, Uchacz T, Sarvas C, Griffiths R, Kuzma M, Wan J, Huang Y. Shoot-specific down-regulation of protein farnesyltransferase (α-subunit) for yield protection against drought in canola. Molecular Plant. 2009; 2(1): 191-200

[17] Wang QB, Xu W, Xue QZ, Su WA. Transgenic *Brassica chinensis* plants expressing a bacterial *codA* gene exhibit enhanced tolerance to extreme temperature and high salinity. Journal of Zhejiang University Science B. 2010; 11(11): 851-861

[18] Seo YJ, Park JB, Cho YJ, Jung C, Seo HS, Park SK, Nahm BH, Song JT. Overexpression of the ethylene-responsive factor gene BrERF4 from *Brassica rapa* increases tolerance to salt and drought in *Arabidopsis* plants. Molecules and Cells. 2010; 30(3): 271-277

[19] Kim JA, Jung HE, Hong JK, Hermand V, McClung CR, Lee YH, Kim JY, Lee SI, Jeong MJ, Kim J, Yun D. Reduction of GIGANTEA expression in transgenic *Brassica rapa* enhances salt tolerance. Plant cell reports. 2016; 35(9): 1943-1954

[20] Maqbool S, Zhong H, El-Maghraby Y, Ahmad A, Chai B, Wang W, Sabzikar R, Sticklen M. Competence of oat (*Avena sativa* L.) shoot apical meristems for integrative transformation, inherited expression, and osmotic tolerance of transgenic lines containing hva1. Theoretical and Applied Genetics. 2002; 105(2-3): 201-218

[21] Agarwal PK, Agarwal P, Reddy MK, Sopory SK. Role of DREB transcription factors in abiotic and biotic stress tolerance in plants. Plant Cell Reports. 2006; 25 (12): 1263-1274

[22] Dalal M, Tayal D, Chinnusamy V, Bansal KC. Abiotic stress and ABA-inducible Group 4 LEA from *Brassica napus* plays a key role in salt and drought tolerance. Journal of Biotechnology. 2009; 139 (2): 137-145

[23] Yang M, Yang Q, Fu T, Zhou Y. Overexpression of the *Brassica napus* BnLAS gene in *Arabidopsis* affects plant development and increases drought tolerance. Plant Cell Reports. 2011; 30(3): 373-388

[24] Lata C, Prasad M. Role of DREBs in regulation of abiotic stress responses in plants. Journal of Experimental Botany. 2011; 62(14): 4731-4748

[25] Luo J, Tang S, Mei F, Peng X, Li J, Li X, Yan X, Zeng X, Liu F, Wu Y, Wu G. BnSIP1-1, a trihelix family gene, mediates abiotic stress tolerance and ABA signaling in *Brassica napus*. Frontiers in Plant Science. 2017; 8: 44

[26] Xiao QS, Zhang XK, Xu BB, Cheng Y, Zheng PY, Lu GY. Cloning and expression pattern of *AnnBn1* gene in *Brassica napus*. Chinese journal of oil crop sciences. 2012; 34: 123-128

[27] Ali N, Zada A, Ali M, Hussain Z. Isolation and identification of *Agrobacterium tumefaciens* from the galls of peach tree. Journal of Pure and Applied Agriculture. 2016; 1(1): 39-48

[28] Shinwari ZK, Nakashima K, Miura S, Kasuga M, Seki M, Yamaguchi-Shinozaki K, Shinozaki K. An *Arabidopsis* gene family encoding DRE/CRT binding proteins involved in low-temperature-responsive gene expression. Biochemical and biophysical research communications. 1998; 250 (1):161-170

[29] Bae MJ, Kim YS, Kim IS, Choe YH, Lee EJ, Kim YH, Park HM, Yoon HS. Transgenic rice overexpressing the *Brassica juncea* gamma-glutamylcysteine synthetase gene enhances tolerance to abiotic stress and improves grain yield under paddy field conditions. Molecular breeding. 2013; 31 (4): 931-945

[30] Saha B, Mishra S, Awasthi JP, Sahoo L, Panda SK. Enhanced drought and salinity tolerance in transgenic

mustard [*Brassica juncea* (L.) Czern&Coss.]overexpressing *Arabidopsis* group 4 late embryogenesis abundant gene (AtLEA4-1). Environmental and Experimental Botany. 2016; 128: 99-111

[31] Rajwanshi R, Kumar D, Yusuf MA, DebRoy S, Sarin NB. Stress-inducible overexpression of glyoxalase I is preferable to its constitutive overexpression for abiotic stress tolerance in transgenic *Brassica juncea*. Molecular Breeding. 2016; 36 (6):76

[32] Ahmed I, Yadav D, Shukla P, Vineeth TV, Sharma PC, Kirti PB. Constitutive expression of *Brassica juncea* annexin, AnnBj2 confers salt tolerance and glucose and ABA insensitivity in mustard transgenic plants. Plant Science. 2017; 265: 12-28

[33] Ren ZH, Gao JP, Li LG, Cai XL, Huang W, Chao DY, Zhu MZ, Wang ZY, Luan S, Lin HX. A rice quantitative trait locus for salt tolerance encodes a sodium transporter. Nature Genetics. 2005; 37(10): 1141-1146

[34] Zhang JZ, Creelman RA, Zhu JK. From laboratory to field. Using information from *Arabidopsis* to engineer salt, cold, and drought tolerance in crops. Plant physiology. 2004; 135(2): 615-621

[35] Farooq S, Farooq-E-Azam. Co-Existence of Salt and Drought Tolerance in *Triticeae*. Hereditas. 2001; 135(2-3): 205-210

[36] Rozema J, Flowers T. Crops for a salinized world. Science. 2008: 1478-1480.

[37] Kaur P, Ghai N, Sangha MK. Induction of thermotolerance through heat acclimation and salicylic acid in *Brassica* species. African Journal of Biotechnology. 2009; 8 (4).

[38] Patidar OP, Yadava DK, Singh N, Saini N, Vasudev S and Yashpal. Deciphering selection criteria for Indian mustard (*Brassica juncea* L.) encountering high temperature stress during post-reproductive phase. International Journal of Chemical Studies. 2020: 8 (4): 2497-2502

[39] Azharudheen TM, Yadava DK, Singh N, Vasudev S, Prabhu KV. Screening Indian mustard [*Brassica juncea* (L.) Czern and Coss)] germplasm for seedling thermo-tolerance using a new screening protocol. African Journal of Agricultural Research. 2013; 8 (38): 4755-4760

[40] Angadi SV, Cutforth HW, Miller PR, McConkey BG, Entz MH, Brandt SA, Volkmar KM. Response of three *Brassica* species to high temperature stress during reproductive growth. Canadian Journal of Plant Science. 2000; 80 (4): 693-701

[41] Young LW, Wilen RW, Bonham-Smith PC. High temperature stress of *Brassica napus* during flowering reduces micro-and megagametophyte fertility, induces fruit abortion, and disrupts seed production. Journal of Experimental Botany. 2004; 55 (396): 485-495

[42] Kumar S, Sairam RK, Prabhu KV. Physiological traits for high temperature stress tolerance in *Brassica juncea*. Indian Journal of Plant Physiology. 2013; 18 (1): 89-93

[43] Chen Z, Wang Z, Yang Y, Li M, Xu B. Abscisic acid and brassinolide combined application synergistically enhances drought tolerance and photosynthesis of tall fescue under water stress. Scientia Horticulturae. 2018; 228: 1-9

[44] Kagale S, Divi UK, Krochko JE, Keller WA, Krishna P. Brassinosteroid confers tolerance in *Arabidopsis thaliana* and *Brassica napus* to a range of abiotic stresses. Planta. 2007; 225 (2): 353-364

[45] Zhang F, Lu K, Gu Y, Zhang L, Li W, Li Z. Effects of low-temperature

stress and brassinolide application on the photosynthesis and leaf structure of tung tree seedlings. Frontiers in Plant Science. 2020; 10: 1767

[46] Marcec MJ, Gilroy S, Poovaiah BW, Tanaka K. Mutual interplay of Ca^{2+} and ROS signaling in plant immune response. Plant Science. 2019; 283: 343-354

[47] Sharma G, Kumar VD, Haque A, Bhat SR, Prakash S, Chopra VL. *Brassica* coenospecies: a rich reservoir for genetic resistance to leaf spot caused by *Alternaria brassicae*. Euphytica. 2002; 125 (3): 411-417

[48] Kumar GO, Chakravarty NV. A simple weather based forewarning model for white rust in *Brassica*. Journal of Agrometeorology. 2008; 10 (1): 75-80

[49] Dev D, Tewari AK, Upadhyay P, Daniel GR. Identification and nomenclature of *Albugo candida* pathotypes of Indian origin causing white rust disease of rapeseed-mustard. European Journal of Plant Pathology. 2020; 158 (4): 987-1004

[50] Chauhan SK, Sharma JB. Inheritance of white rust resistance in Indian mustard incorporated from *Brassica napus*. Indian J. Genet. 2001; 61 (3): 250-252

[51] Ramos MJ, Fernández CM, Casas A, Rodríguez L, Pérez Á. Influence of fatty acid composition of raw materials on biodiesel properties. Bioresource Technology. 2009; 100 (1): 261-268

[52] Yoshie-Stark Y, Wada Y, Wäsche A. Chemical composition, functional properties, and bioactivities of rapeseed protein isolates. Food Chemistry. 2008; 107 (1): 32-39

[53] Love HK, Rakow G, Raney JP, Downey RK. Development of low glucosinolate mustard. Canadian Journal of Plant Science. 1990; 70 (2): 419-424

[54] Love HK, Rakow G, Raney JP, Downey RK. Genetic control of 2-propenyl and 3-butenyl glucosinolate synthesis in mustard. Canadian Journal of Plant Science. 1990; 70(2): 425-429

[55] Sodhi YS, Mukhopadhyay A, Arumugam N, Verma JK, Gupta V, Pental D, Pradhan AK. Genetic analysis of total glucosinolate in crosses involving a high glucosinolate Indian variety and a low glucosinolate line of *Brassica juncea*. Plant breeding. 2002; 121 (6): 508-511

[56] Mahmood T, Ekuere U, Yeh F, Good AG, Stringam GR. Molecular mapping of seed aliphatic glucosinolates in *Brassica juncea*. Genome. 2003; 46 (5): 753-760.

[57] Sun FJ. Hybrid vigor in *Brassica*.J AgricAssoc China. 1943; 175: 35-58

[58] Ogura H. Studies on the new male sterility in Japanese radish, with special references on the utilization of this sterility towards the practical raising of hybrid seeds.Mem. Fac. Agric. Kagoshima Univ. 1968; 6: 40-75

[59] Rawat DS, Anand IJ. Male sterility in Indian mustard. Indian J Genet Plant Breed. 1979; 39: 412-414

[60] Rao GU, Batra-Sarup V, Prakash S, Shivanna KR. Development of a New Cytoplasmic Male-Sterility System in *Brassica juncea* through Wide Hybridization. Plant breeding. 1994; 112 (2): 171-174

[61] Kirti PB, Mohapatra T, Khanna H, Prakash S, Chopra VL. *Diplotaxis catholica* + *Brassica juncea* somatic hybrids: molecular and cytogenetic characterization. Plant cell reports. 1995; 14 (9): 593-597

[62] Prakash S, Kirti PB, Chopra VL. Cytoplasmic male sterility (CMS) systems other than *ogu* and *polima* in *Brassica*: current status. In Proc 9th Int Rapeseed Conf. 1995; 1: 44-48

[63] Kirti PB. Development of a stable cytoplasmic male sterile line of *Brassica*

juncea from somatic hybrid *Trachystoma ballii* + *Brassica juncea*. Plant Breed.. 1995; 114: 434-438

[64] Banga SS, Banga SK. *Enarthrocarpus lyratus* cytoplasm causes male sterility in oilseed rape. In International Symposium on Heterosis in Crops, Mexico City. 1997; 17-22

[65] Prakash S, Ahuja I, Upreti HC, Kumar VD, Bhat SR, Kirti PB, Chopra VL. Expression of male sterility in alloplasmic Brassica *juncea* with *Eruca strumcanariense* cytoplasm and the development of a fertility restoration system. Plant Breeding. 2001; 120 (6): 479-482

[66] Bhat SR, Vijayan P, Ashutosh, Dwivedi KK, Prakash S. *Diplotaxis erucoides*-induced cytoplasmic male sterility in *Brassica juncea* is rescued by the *Moricandia arvensis* restorer: genetic and molecular analyses. Plant breeding. 2006; 125 (2): 150-155

[67] Sodhi YS, Chandra A, Verma JK, Arumugam N, Mukhopadhyay A, Gupta V, Pental D, Pradhan AK. A new cytoplasmic male sterility system for hybrid seed production in Indian oilseed mustard *Brassica juncea*.Theoretical and Applied Genetics. 2006; 114 (1): 93

[68] Bhat SR, Kumar P, Prakash S. An improved cytoplasmic male sterile (*Diplotaxis berthautii*) *Brassica juncea*: identification of restorer and molecular characterization. Euphytica. 2008; 159(1-2): 145-152

[69] Jagannath A, Arumugam N, Gupta V, Pradhan A, Burma PK, Pental D. Development of transgenic barstar lines and identification of a male sterile (barnase)/restorer (barstar) combination for heterosis breeding in Indian oilseed mustard (*Brassica juncea*). Current Science. 2002; 82 (1): 46-51

[70] Chand S, Patidar OM, Meena VK, Shiv A. Barnase-barstar system: an indelible technique to produce hybrid seeds in self-pollinated crops. International Journal of Farm Sciences. 2018; 8 (2): 109-113

[71] Singh D, Mehta R. Studies on breeding brown sarson. I. Comparison of F1's and their parents. Indian J. Genet. Pl. Breed. 1954; 14: 74-77

[72] Chand S, Chandra K, Khatik CL. Varietal Release, Notification and Denotification System in India. In Plant Breeding-Current and Future Views. 2020. IntechOpen

[73] Prakash S, Chopra VL. Synthesis of alloplasmic *Brassica campestris* as a new source of cytoplasmic male sterility. Plant breeding. 1988; 101 (3): 253

[74] Kirti PB, Narasimhulu SB, Mohapatra T, Prakash S, Chopra VL. Correction of chlorophyll deficiency in alloplasmic male sterile *Brassica juncea* through recombination between chloroplast genomes. Genetics Research. 1993; 62 (1): 11-14

[75] Kumawat G, Kumawat CK, Chandra K, Pandey S, Chand S, Mishra UN, Lenka D, Sharma R. Insights into Marker Assisted Selection and Its Applications in Plant Breeding. In Plant Breeding-Current and Future Views. 2020; IntechOpen

[76] Sharma SK and Singh R. Genetic resources of oilseed crops in India. In: Hegde DM (Ed).Changing Global Vegetable Oils Scenario: Issues and Challenges before India. Indian Society of Oilseed Research, DOR, Hyderabad. 2007; 1-16

[77] Watts A, Sankaranarayanan S, Raipuria RK, Watts A. Production and Application of Doubled Haploid in *Brassica* Improvement. In Brassica Improvement Springer, Cham. 2020; 67-84

[78] Jain RK, Sharma DR, Chowdhury JB. High frequency regeneration and

heritable somaclonal variation in *Brassica juncea*. Euphytica. 1989; 40 (1-2): 75-81

[79] Anuradha G, Narasimhulu SB, Arunachalam V, Chopra VL. A comparative evaluation of somaclonal, gamma ray and EMS induced variation in *Brassica juncea*. Journal of Plant Biochemistry and Biotechnology. 1992; 1 (2): 105-108

[80] Walden R, Koncz C, Schell J. The use of gene vectors in plant molecular biology. Methods Mol. Cell. Biol. 1990; 1: 175-194

[81] Mukherjee AK, Mohapatra T, Varshney A, Sharma R, Sharma RP. Molecular mapping of a locus controlling resistance to *Albugo candida* in Indian mustard. Plant Breeding. 2001; 120 (6): 483-497

[82] Padmaja KL, Arumugam N, Gupta V, Mukhopadhyay A, Sodhi YS, Pental D, Pradhan AK. Mapping and tagging of seed coat colour and the identification of microsatellite markers for marker-assisted manipulation of the trait in *Brassica juncea*. Theoretical and Applied Genetics. 2005; 111 (1): 8-14

[83] Liu S, Huang H, Yi X, Zhang Y, Yang Q, Zhang C, Fan C, Zhou Y. Dissection of genetic architecture for glucosinolate accumulations in leaves and seeds of *Brassica napus* by genome-wide association study. Plant Biotechnology Journal. 2020; 18 (6): 1472-1484

[84] Kaur J, Akhatar J, Goyal A, Kaur N, Kaur S, Mittal M, Kumar N, Sharma H, Banga S, Banga SS. Genome wide association mapping and candidate gene analysis for pod shatter resistance in *Brassica juncea* and its progenitor species. Molecular Biology Reports. 2020; 1-2

[85] Cai G, Yang Q, Yang Q, Zhao Z, Chen H, Wu J, Fan C, Zhou Y. Identification of candidate genes of QTLs for seed weight in *Brassica napus* through comparative mapping among *Arabidopsis* and *Brassica* species. BMC genetics. 2012; 13 (1): 105

[86] Bisht NC, Gupta V, Ramchiary N, Sodhi YS, Mukhopadhyay A, Arumugam N, Pental D, Pradhan AK. Fine mapping of loci involved with glucosinolate biosynthesis in oilseed mustard (*Brassica juncea*) using genomic information from allied species. Theoretical and Applied Genetics. 2009; 118 (3): 413-421

[87] Couvreur TL, Franzke A, Al-Shehbaz IA, Bakker FT, Koch MA, Mummenhoff K. Molecular phylogenetics, temporal diversification, and principles of evolution in the mustard family (Brassicaceae). MolBiolEvol. 2010; 27 (1): 55-71

[88] Augustine R, Majee M, Pradhan AK, Bisht NC. Genomic origin, expression differentiation and regulation of multiple genes encoding CYP83A1, a key enzyme for core glucosinolate biosynthesis, from the allotetraploid *Brassica juncea*. Planta. 2015; 241 (3): 651-665

[89] Heng S, Gao J, Wei C, Chen F, Li X, Wen J, Yi B, Ma C, Tu J, Fu T, Shen J. Transcript levels of orf288 are associated with the hau cytoplasmic male sterility system and altered nuclear gene expression in *Brassica juncea*. Journal of experimental botany. 2018; 69 (3): 455-466

[90] Bhattacharya S, Das N, Maiti MK. Cumulative effect of heterologous AtWRI1 gene expression and endogenous BjAGPase gene silencing increases seed lipid content in Indian mustard *Brassica juncea*. Plant Physiology and Biochemistry. 2016; 107: 204-213

[91] Savadi S, Naresh V, Kumar V, Bhat SR. Seed-specific overexpression of Arabidopsi s DGAT1 in Indian mustard (*Brassica juncea*) increases seed oil content and seed weight. Botany. 2016; 94 (3): 177-184

[92] Zhao N, Xu X, Wamboldt Y, Mackenzie SA, Yang X, Hu Z, Yang J, Zhang M. MutS HOMOLOG1 silencing mediates ORF220 substoichiometric shifting and causes male sterility in *Brassica juncea.* Journal of experimental botany. 2016; 67 (1): 435-444

[93] Zheng M, Zhang L, Tang M, Liu J, Liu H, Yang H, Fan S, Terzaghi W, Wang H, Hua W. Knockout of two Bna MAX 1 homologs by CRISPR/ Cas9-targeted mutagenesis improves plant architecture and increases yield in rapeseed (*Brassica napus* L.). Plant biotechnology journal. 2020; 18 (3): 644-654

[94] Mihr C, Baumgärtner M, Dieterich JH, Schmitz UK, Braun HP. Proteomic approach for investigation of cytoplasmic male sterility (CMS) in *Brassica*.Journal of plant physiology. 2001; 158 (6): 787-794

[95] Mohammadi PP, Moieni A, Komatsu S. Comparative proteome analysis of drought-sensitive and drought-tolerant rapeseed roots and their hybrid F1 line under drought stress. Amino Acids. 2012; 43 (5): 2137-2152

[96] Yousuf PY, Ahmad A, Ganie AH, Iqbal M. Salt stress-induced modulations in the shoot proteome of *Brassica juncea* genotypes. Environmental Science and Pollution Research. 2016a; 23 (3): 2391-2401

[97] Yousuf PY, Ganie AH, Khan I, Qureshi MI, Ibrahim MM, Sarwat M, Iqbal M, Ahmad A. Nitrogen-efficient and nitrogen-inefficient Indian mustard showed differential expression pattern of proteins in response to elevated CO2 and low nitrogen. Frontiers in plant science. 2016b; 7: 1074

[98] Sinha S, Jha JK, Maiti MK, Basu A, Mukhopadhyay UK, Sen SK. Metabolic engineering of fatty acid biosynthesis in Indian mustard (*Brassica juncea*) improves nutritional quality of seed oil. Plant Biotechnology Reports. 2007; 1 (4): 185-197

[99] Kortesniemi M, Vuorinen AL, Sinkkonen J, Yang B, Rajala A, Kallio H. NMR metabolomics of ripened and developing oilseed rape (*Brassica napus*) and turnip rape (*Brassica rapa*). Food chemistry. 2015; 172: 63-70

[100] ICAR-Directorate of Rapeseed-Mustard Research, https://www.drmr.res.in accessed on 07.01.2021

Chapter 3

Innovative Strategies to Develop Abiotic and Biotic Stress Tolerance in Mustard (Brassicaceae)

Bahaderjeet Singh, Amanpreet Singh Sran and Gagandeep Singh Sohi

Abstract

Mustard crop is the third important source of vegetable oil randomly below soybean *L.* and palm, all over the world. Brassica crop is extremely susceptible to some biotic and abiotic stresses and they significantly influence the quality and quantity of the crop. In the past generally breeding techniques are used to develop resistance in mustard to avoid diseases though various pathogens are soon able to overcome that resistance by modifying their metabolic cycles. To bear the challenge there is an urgent need to develop abiotic as well as biotic stress tolerant plants using advanced techniques by understanding metabolic and biochemical pathways of plants and pathogens. Several techniques such selection of stress tolerance microbes, metabolite, enzymes, and genes are very important to avoid stresses. Whereas several techniques such as deployment of molecular markers for breeding, identification of Quantitative trait loci (QTL), *in vitro* tissue culture etc. can be more useful to improve biotic and abiotic stress tolerance in mustard. To develop healthy and high yield varieties, the mix of these techniques is needs to be implemented.

Keywords: stress tolerance, mustard, molecular approaches, biological approaches

1. Introduction

Rapeseed-mustard is the third most important oilseed crop after soybean and groundnut, contributing nearly about 20–25% of the total oilseed production in the country. In India, rapeseed-mustard occupy approximately 22.2% of total oilseeds cultivated area and approximately 32% of the country's total oilseed production, with an area of 6.00mha, production of 8.04mt and yield of 1339 kg/ha during 2017–2018 [1]. Brassica oilseed crops play a vital role in the diversification in cropping system and also in providing the quality food by meeting the fat requirement to same extent. Mustard is considered to be of high economic importance in local as well as international trade and it is one of the major contributors to Yellow Revolution. Many biotic and abiotic stresses are liable for reducing the production and productivity of rapeseed-mustard. Biotic stresses are diseases and pests and abiotic stresses are due to environmental factors like temperature, salinity, drought, frost and Water logging stress. Diseases play a pivotal role in reducing the quality and quantity of mustard crop [2, 3]. Unfavorable environmental conditions severely affect growth, productivity and genome stability of the crop. These unsuitable ecological factors are a hazardous for plants that avert them from reaching their full genetic potential and decrease the crop productivity worldwide. Various abiotic Stresses, such as extreme temperature, salinity and heavy metal toxicity cause massive crop yield loss. Pattern of climate is becoming more erratic globally with increased occurrence of drought, flood, storms, heat waves, and seawater intrusion.

According to Howe and Jander [4], array of morphological, genetic, biochemical and molecular processes has been targeted to develop resistance in plants. These mechanisms may be expressed constitutively as preformed resistance, or they may be inducible and deployed only after attack. The latest studies indicate that the plant mechanisms of disease resistance or susceptibility are related to mechanistic response [5]. Various plant pattern recognition receptors (PRRs) that sense pathogens or conserved molecules termed pathogen-associated molecular patterns (PAMPs) and then induce PAMP triggered immunity (PTI), in case of biotic stress. While in case of abiotic stress plants respond to various stress factors such as salinity, heat, cold, drought, excess water, heavy metal toxicity, nutrient loss and pass information through multifaceted molecular signaling pathways leading to expression of stress-related genes. These responses at the molecular, cellular, physiological and biochemical levels enable the plants to survive [6].

The new biological and molecular tools have been opened up new perspectives in stress biology and can be applied in Mustard to develop biotic and abiotic stress tolerance. The omics approaches such as genomics, proteomics, metabolomics and transcriptomics have direct potential for improving stress tolerance in plants. The use of PGPRs, beneficial metabolites and enzymes produced by microorganisms are found to effective in biotic and abiotic stress management in mustard.

2. Abiotic stresses in mustard and strategies to develop tolerance

Productivity of brassica crop is affected by a various abiotic stresses. These may include deficit or excess water availability, salinity levels in soil as well as in irrigation water and extreme temperatures. In addition, mineral deficiency or toxicity and excessive chemical content in soil are frequently faced by plants. Sometimes various abiotic stresses occurs in combination and affect the plants severely. For example, scarcity of water and high temperature are commonly occurs in the period of drought and can be induced by mineral toxicities that restrict root growth. Further, plants are also exposed to salinity, drought and frost-like conditions in combination in many cases. Abiotic stresses are primarily unavoidable and are the most harmful factor concerning the growth and productivity of brassica crops.

2.1 Strategies to develop salinity stress tolerance in mustard

Salt stress is one of the major limiting factor that disturb the yield and other agronomic important characters of mustard. Vital [7] observed that, soil texture and composition adversely affected by one of the major environmental factor that is salt stress. High salt concentration leads to imbalance of nutrients and ions, it reduces the normal morpho-physiological and other biological processes of mustard [8]. High salt concentration negatively affects the seed germination in many Brassica species, also showed retardation in plant growth and development, resulting in reduced crop yield and even death of plant under severe conditions [9]. Due to high concentration of salt content, the osmotic pressure of soil is higher than the root cells, thereby root cells instead of absorbing water from soil lose water leading to water and nutrition imbalance in plants, thus adversely affects plant growth [10]. Plants use different resources to sense, counter and acclimatize to altering the saline environment based by making modifications in morphological, physiological traits and molecular metabolism which may further be enhanced by thiourea TU induction. Recently published reports have thoroughly explained the function of TU in inducing the salt tolerance and primary mechanisms in many plants, including Indian mustard (*Brassica juncea*) [11]. Evidence suggests that TU treatment (6.5 mM) improved salt tolerance in *Brassica juncea* by enhancing the translocation of sucrose from source to sink [12]. Recently, it has been discovered that mitochondria play a critical role in plant protection again salinity stress [13]. This is an important mechanism by which TU maintains mitochondrial homeostasis and ATPases (FoF1-ATP synthase) plays an important role in TU-induced salt tolerance in *Brassica juncea* [14]. TU application can also alleviate the adverse effects of salt stress by inducing changes in transcription through the modulation of microRNA and hormone production [15]. A prolific root system is important to improve stress tolerance and final yield [16]. Endophytic *P. indica* induces salt tolerance in mustard by increasing the levels of antioxidants. The continous exposure of 500 mM NaCl solution nonsymbiotic plants *Leymus mollis* (dunegrass) cause severe wilting and desiccation in 7 days and the plants were dead after 14 days. Contrary to this the symbiotic plants infected with *Fusarium culmorum* did not show signs of wilting even exposed to 500 mM NaCl solution for 14 days [17]. Salt tolerant varieties like CS52, CS54, CS56, CS58, CS 234–4 and Narendra Rai have better tolerance potential can be grown in such condition.

2.2 Strategies to develop Drought stress tolerance in mustard

Drought can severely affect seed traits such as seed germination, seed yield and seed quality as well as plant vegetative growth. Shekari et al. reported that the most sensitive stage for drought injury was flowering resulting in high loss in seed as well as oil yield by 29.5% and 31.7%, respectively [18]. Hasanuzzaman et al. examined that *Brassica napus* may be more resistant to drought stress than that of *Brassica rapa* [19].

The challenge is even greater for developing drought tolerant trait in plants for water-limited environments where occurrence, timing and severity of drought may fluctuate from one zone to the next and also over the years. Furthermore, it induces large impacts on emergence, growth, quantity and quality of produce production through phenological, physiological and biochemical pathways [20]. Physiological changes in water potential and relative water content of the water-stressed leaves through osmoregulation and osmotic adjustment have been observed in Brassica crops. In the process of physiological adaptation, maintenance of turgor pressure appears to be the central process. In crop production, different techniques are used to conserve water and increase water use efficiency in order to tackle water scarcity. One of them is planting method which affects the plant population and nutrient availability. Most commonly used cheapest method for water conservation are drill sowing raised bed planting and furrow planting. Seedling establishment is a phonological stage at which drought stress could be damaging. Seed broadcasting technique of Brassica crops results in uneven distribution and leads to imbalances availability of water, space and nutrients and poor seedling establishment and ultimately lower yield. Plant growth regulators (PGRs) regulate the germination, formation and distortion of roots, leaves and stem elongation and ripening etc. exogenous application of salicylic acid, gibbereallic acid and cytokinins also improves stress tolerance in mustard. Potassium (K) is one of the key plant nutrients and is involved in drought mitigation by regulating turgor pressure, photosynthesis, translocation of assimilates to various organs and enzyme activation [21].

Other than traditional techniques the Transcription factors (TFs) are emerging as useful resources for genetic engineering to induce drought tolerance in mustard plants, because they act as mjors regulators of various stress-regulatory pathways. Many TFs belonging to families AP2/EREBP, MYB, WRKY, NAC, bZIP have been involved in drought stress tolerance and some TF genes have also been engineered to develop stress resistance in plants. TFs are very important regulators, as they function as terminal transducers and comprehensively regulate the expression of group of downstream genes by combination of specific cis elements in their promoter region [22]. Over-expression of a constitutively active form of AtDREB2A from Arabidopsis has been reported to improve the tolerance to drought and osmotic stresses [23].

2.3 Strategies to develop cold stress tolerance in mustard

Frost is a sudden crop killer with devastating threat, especially in the north and northeastern parts of India where temperature unexpectedly drops below 0°C. Low-temperature stress not only decreases grain yield but also affects crop grain quality [24]. Shah et al. reported that whole plant death in mustard if frost stress affects the seedling stages. The injury rate of frost stress depends on many important components such as duration and amount of cold stress, different stages of plant growth and moisture content. It has direct effect on the flowering and siliqua development and prevents seed formation, thereby affecting crop productivity, causing considerable yield loss [25].

Cold stress tolerance mechanism in plants is regulated via transcriptional activation or repression. The majority of stress associated proteins such as heat shock, chemical shock and late embryogenesis abundant proteins (HSPs, CSPs and LEA) are accumulate upon extreme temperature stress. They act as molecular chaperones, which are responsible for protecting the cellular machinery in a broad range of cellular processes. Evidence suggests that cold tolerance is linked with the increased expression of genes involved in transcriptional regulation, osmotic adjustment, antioxidant defense and metabolite biosynthesis [26]. Varieties like RGN-48, RK-9001, RH-8816, RGN-13, RH-819, Swaranjyoti, RH-781 have been reported to have good frost tolerance. Mustard crop can be also protected from frost by chemical spray of dimethyl sulphoxide, dithane or 0.15% of H_2SO_4.

2.4 Strategies to develop heavy metal toxicity stress tolerance in mustard

Plants largely depend on soil solution to acquire nutrients for their growth and developmental cycle. The recent increase in contamination of arable lands with heavy metals is one of the most important causes of loss in crop productivity [27]. Extensive exposure to heavy metal contamination threatens the sustainability of environmental and agricultural systems. Crops are routinely subjected to metal toxicity due to improper irrigation methods and the addition of excessive quantities of chemical fertilizers, and other synthetic nutrients [28]. Some (potentially toxic) heavy metals, such as Cu, Zn, Ni, Co, Se, and Fe, are also essential elements required for the optimal performance of plants and become toxic when accumulated in excess in soil solution [29, 30]. On the other hand, non-essential elements, such as arsenate (As), cesium (Cs), lead (Pb), and cadmium (Cd), can hamper crop productivity when accumulated in the soil even in trace amounts [31]. Soil contamination with heavy metals causes accumulation of these toxic metals in plant parts, resulting in decreased crop productivity and increased risk to animal and human health [32].

The root-associated dark septate endophyte (DSE), *Exophiala pisciphila* isolated from *Zea mays* showed enhanced antioxidant enzyme activity under increased soil Cadmimium (Cd) stress [33]. Three important genes involved in uptake, detoxification and transport of Cd were recognized as downregulation of ZIP, upregulation of PCS and MTP upon inoculation with DSE and exposed to high concentration of Cd. The *Pseudomonas* and *Gigaspora* are reported to alter level of 1-aminocyclopropane-1-carboxylate (ACC) which further increase the tolerance of heavy metals by directly manipulating the ethylene levels in plants [34].

2.5 Identification of quantitative trait loci (QTL) to develop abiotic stress tolerance in mustard

Quantitative trait locus (QTL) is a statistical method developed to analyse and correlate the phenotypic and genotypic data to estimate the genetic variations in complex traits. This technique is less time consuming and gives better mapping resolution by exploring and utilizing each event of recombination that occurs in the evolutionary history. In various crops, QTLs were recognized for a several beneficial agronomic traits, such as enhancing abiotic and biotic stress tolerance, yield and yield contributing factors of the crop, flowering time, root development and uptake of nutrients and nitrogen fixation. Molecular markers linked with various agronomic traits derived from association mapping are reported in crops including soybean [35] and brassica [36, 37]. Lu et al. reported that resequencing of 588 *Brassica napus* accessions from 21 countries has generated 5,294,158 (single nucleotide polymorphisms) SNPsand 1,307,151 indels. The genome-wide association study (GWAS) find 60 loci considerably associated with agronomic traits such as stress tolerance, seed quality etc., which may be proven as a valuable resource for genetic improvement [38].

All these factors such as salt stress, drought and frost decline the crop quality and quantity. These abiotic stresses can be managed by exploring genetic resources material and agronomic factors. Donor lines have been identified for different abiotic stresses and are being used in the breeding programmes for developing tolerant varieties against abiotic stresses. A number of improved cultivars or hybrids have been developed which perform better under different type of stresses resulting in lower reduction upon exposure to stress compared to high yielding varieties. Conventional techniques of plant breeding have not been proved that much successful in addressing abiotic stresses mitigation so far. Therefore reason and need for adoption of new molecular approaches.

3. Biotic stresses in mustard and strategies to develop tolerance

Besides environmental stresses, biotic stresses are diseases and insect-pests. The fungal diseases are considered as an important biotic constraint in mustard, which leads to significant yield losses of crop world-wide. More than thirty diseases are known to occur on brassica crops in India [39]. However, only few of them are considered as major diseases on the basis of economic yield losses and according to their distribution in the country. Major biotic stresses of rapeseed-mustard in India are Alternaria blight [*Alternaria brassicae* (Berk.) Sacc.], white rust [*Albugo candida* (lev.) Kuntze], Powdery mildew (*Erysiphe cruciferarum*), Downy mildew (Hyaloperonospora parasitica) and Sclerotinia rot (*Sclerotinia sclerotiorum*) which influence the quality and quantity of seed [40–42]. Alternaria blight [*A. brassicae* (Berk.) Sacc.] and White rust [*A. candida* (lev.) Kuntze] have been reported to be most wide spread and destructive fungus diseases of mustard all over the world [43]. The details of these diseases and their causal organism which are affecting rapeseed-mustard crop in India are mentioned in **Table 1** below.

3.1 Conventional approaches for disease management

The following strategies are helpful for disease management.

- Optimization of sowing time is very important as it significantly affects the disease incidence and severity.
- Use of certified seeds, chemicals, disease resistant and tolerant varieties.
- Seed treatment with various biocontrol agents viz., *T. viride*, *G. virens*, *Pseudomonas* or botanicals like *Allium sativum* etc. Use of bio-control agents (BCA) is advantageous as they are often effective against a wide range of soil-borne pathogens. Moreover, they are ecofriendly, cost effective and their use avoids the risk of development of resistance in the pathogen towards the control agent.
- Use recommended doses of N, P and K fertilizers, maintaining optimum plant population with recommended crop spacing.

Application of more use of pesticides leads to the development of resistance in the target pests, and has negative impacts on biodiversity. More importantly, plants will have increased susceptibility to pests due to the implications of changes of climate. Under such conditions, gaining better knowledge on physiology of plants could lead to sustainable control of biotic stresses. Several plant breeding techniques has been used in extensively to develop biotic stress tolerance in mustard. The stress responses in plants is showed high levels of complexity and redundancy at the sensitivity, response and expression levels with interconnection between stress pathways and over lapping functions between stress metabolites and stress proteins in different stresses. In the case of stress proteins, there are limits on genes of known function that are available but perhaps more importantly the issue of whether single or multiple gene transformations will confer stable resistance. Regular upgradation of technology is required to develop better solution of biotic stresses. Now-a-days, several molecular techniques are considered for a better crop disease management. Lack of disease resistance sources of plant breeding is serious problem and difficult challenge for crop improvement and this problem can be solved through using biotechnology approaches. This is one of the best option and

Disease	Causal organism	Yield losses
Alternaria blight	*Alternaria. brassicae* (Berk.) Sacc.	10–70
White rust	*Albugo candida* (lev.) Kuntze	Upto 47
Sclerotinia rot	*Sclerotinia sclerotiorum*	Upto 35
Downy mildew	*Hyaloperonospora parasitica*	17—37
Powdery mildew	*Erysiphe cruciferarum*	Upto 18

Source: Directorate of Rapeseed-Mustard Research (DRMR).

Table 1.
Economically important biotic stresses (diseases) of mustard and losses caused by them.

opportunity to develop strategies for biotic stress tolerance in crop [44]. Several methods has been employed to develop biotic stress tolerance are discussed below.

3.2 Molecular approaches for disease management

3.2.1 In-vitro tissue culture to develop biotic stress tolerance

The conventional breeding techniques are used for the incorporation of genes of interest from inter-crossing species into the crop for the development of biotic stress tolerance; however, these methods proved less effective with undesirable results [45]. Moreover, biotechnological techniques can be effective for the development of stress-tolerant plants. The genetic transformation involves the transfer of stress tolerance gene from gene pools in various plant species for establishment of stress tolerant crops. Genetic engineering could be most effective for the improvement of crop varieties; however, the major trouble associated with this approach is the low transformation competence, silencing of transgene and low gene expression [46]. Recently, tissue culture has proved to be an appropriate and less costly technique for development of stress-tolerant plants. The tissue culture plants are grown in controlled lab conditions requires limited time and space with potential to develop of stress-tolerant plants and leads to the better understanding of biochemical and metabolic pathways of plants growing in harsh environmental conditions [47]. Using partially purified culture filtrates, *B. napus* showing resistance to *Alternaria brassicicola* has been obtained [48] and *B. napus* showing resistance to *Phoma lingam* has been obtained through embryonic culture [49].

In vitro selection through enhanced expression of pathogenesis-related (PR) proteins, antifungal peptides or biosynthesis of phytoalexins is an important tool for desirable plant selection [50, 51]. This technology is having an upper hand over transgenic approach for developing improved disease-tolerant crops [52]. Developing pathogen resistance through in vitro selection can be carried out using organogenic or embryogenic calli, shoots, somatic embryos or cell suspensions. By exposing these cultures to different toxins produced by various plant pathogens, tolerant plants can be raised [51]. "Pusa Jaikisan" is the first high yielding variety of mustard though tissue culture and suited for nontraditional areas.

3.2.2 RNAi-mediated plant defence to develop biotic stress tolerance

RNA interference (RNAi), is a powerful technology for discovering the functional genetic sequences and harness the down regulation of expression of gene(s) specifically. To accomplish the modified gene expression for a particular trait, gene silencing viz. cosuppression, post transcriptional gene silencing, virus-induced gene silencing etc. can be used. This molecular phenomenon has become a focal point of modern plant biology research across the globe. Thus it has been remarkably used in crop improvement likewise has become a valuable tool for functional genomics in Brassica (Brassica sp.). The rapid adoption of RNAi has replaced previous antisense technology. RNAi has aided in identification of different functions and biological roles of various mustard genes, which are involved in fertility and somatic embryogenesis, resistance to biotic and abiotic stresses and qualitative improvements in oil seed as well as it also have major role in yield and maturity traits.

3.2.3 Use of microorganisms in biotic stress tolerance

The symbiotic interactions between plant and microorganisms may result in several outcomes as defined by fitness benefits by each of the partners [53].

Interaction to host plants can be positive, neutral or negative. Variations in the outside environment put the plant metabolism out of homeostasis, which creates necessity for the plant to harbour some advanced genetic and metabolic mechanisms within its cellular system [54]. The priming of host response against pathogen is termed as induced systemic resistance, in this case the host response is activated by nonpathogenic plant-associated microorganisms. The ISR induce plant defense mechanisms and protects unexposed part of plants against a future attack by pathogenic microbes and insect pests. Plant hormones ethylene and jasmonic acid plays a regulatory function in the network of interrelated signaling pathways involved in ISR induction [55].

Many studies have been dedicated to the induced systemic resistance SR mediated by free-living rhizobacterial strains [56] the resistance in Mustard (*Brassica juncea*) Induced against Alternaria Black Spot using a virulent *Alternaria brassicae* Isolate-D whereas the strains of *Alternaria alternata* failed to induce resistance against *Alternaria brassicae* [57]. Subsequently attention was drawn to ISR mediated by several other species of genus *Pseudomonas* and the effect was characterized in different plant–pathogen systems. *Pseudomonas* sp. Strain-1 was shown to suppress Sclerotinia stem rot incited by *Sclerotinia sclerotiorum* on stem of mustard. Configuring the functions of endophytes there role in stress tolerance increases immensely. Endophytic microbes improves the plant health by deterring herbivory and pathogenesis while also facilitating plant growth through nutrient uptake, water use efficiency and curtailing of environmental stresses. The endophytic bacteria *Pseudomonas syringae* bacterium was able to induce disease resistance via defense priming [58].

4. Conclusion

Based on the foregoing chapter, there is no doubt biotic and abiotic stresses are the major barriers in enhancing the productivity of rapeseed-mustard crop. Conventional plant breeding techniques has not provided us full proof protection against abiotic and biotic stresses so far. So, there is an urgent need to minimize the adverse effects of these stresses on the brassica crops to enhance the productivity and production to meet the ever-growing demand of oil in the country. The use of genetic and genomic analysis which helps to identify DNA regions tightly linked to agronomical traits in mustard. Molecular markers for the indirect selection of improved crops speeds up the selection process by alleviating time-consuming approaches direct screening under screen house and field conditions. Use of PGPRs in mustard can also be highly effective as these microorganisms are able to reduce abiotic as well as biotic stress in plants by producing beneficial enzymes, proteins, hormones etc. Tissue culture is another advantageous technique in mustard as this crop is highly susceptible stresses at initial stages. Therefore the use of one particular technique or the mix of these techniques has a potential to improve the stress tolerance in rapeseed mustard.

Author details

Bahaderjeet Singh*, Amanpreet Singh Sran and Gagandeep Singh Sohi
Guru Kashi University, Bathinda, Punjab, India

*Address all correspondence to: sidhubahaderjit@yahoo.in

References

[1] Anonymous. 2018; https://www.nfsm.gov.in/ReadyReckoner/Oilseeds/Stat_OS 2018.pdf.

[2] Hossain MD, Ali MMH, Ahmed HU. Disease of mustard and groundnut and their management. Proc. of the 2nd National Workshop on oilseed held on 26-29 April, 1992 at BARC, BD.

[3] Prasad L, Vishnuvat K. Assessment of yield loss in cauliflower seed crop due to Alternaria blight. Indian Phytopathology. 2006;59:185-189.

[4] Howe GA, Jander G. Plant immunity to insect herbivores. Annual Reviews in Plant Biology. 2008;59:41-66. DOI: 10.1146/annurev.arplant.59.032607.092825

[5] Ausubel FM. Are innate immune signaling pathways in plants and animals con- served? Nature Immunology. 2005; 6:973-979. DOI: 10.1038/ni1253

[6] Monaghan J, Zipfel C. Plant pattern recognition receptor complexes at the plasma membrane. Current Opinion in Plant Biology. 2012;15(4):349-357. DOI: 10.1016/j.pbi. 2012.05.006

[7] Vital SA, Fowler RW, Virgen A, Gossett DR, Banks SW, Rodriguez J. Opposing roles for superoxide and nitric oxide in the NaCl stress induce regulation of antioxidant enzyme activity in cotton callus tissue. Environmental and Experimental Botany. 2008;62:60-68.

[8] Shinwari ZK, Nakashima K, Miura S, Kasuga M, Seki M, Yamaguchi-Shinozaki K, Shinozaki K. An Arabidopsis gene family encoding DRE/CRT binding proteins involved in low-temperature-responsive gene expression. Biochemical and Biophysical Research Communications. 1998;250(1):161-170.

[9] Zamani Z, Nezami MT, Habibi D, Khorshidi MB. Effect of quantitative and qualitative performance of four canola cultivars (*Brassica napus* L.) to salinity conditions. Advances in Environmental Biology. 2010;4(3):422-427.

[10] Sharma P, Kannu P, Sardana V, Choudhary OP, Banga SS. Physiological and biochemical basis of salinity tolerance in Indian mustard (*B. juncea*). Abstract number: 171. 15th International Rapeseed Congress held at Berlin on June 16-19, 2019.

[11] Srivastava, A. K., Ramaswamy, N. K., Mukopadhyaya, R., Jincy, M. G. & D'Souza, S. F. Thiourea modulates the expression and activity profile of mtATPase under salinity stress in seeds of *Brassica juncea*. Annals of Botany. 2009;103: 403-410. doi: 10.1093/aob/mcn229 (2009).

[12] Srivastava AK, Nathawat NS, Ramaswamy NK, Sahu MP, Singh G. Evidence for thiol-induced enhanced in situ translocation of ^{14}C-sucrose from source to sink in *Brassica juncea*. Environmental and Experimental Botany. 2008; 64: 250-255.

[13] Paiva ALS, Passaia G, Lobo AKM, Jardim-Messeder D, Silveira JAG, Margis-Pinheiro M. Mitochondrial glutathione peroxidase (OsGPX3) has a crucial role in rice protection against salt stress. Environmental and Experimental Botany. 2019;158:12-21.

[14] Srivastava AK, Ramaswamy NK, Mukopadhyaya R, Jincy MG, D'Souza SF. Thiourea modulates the expression and activity profile of mtATPase under salinity stress in seeds of *Brassica juncea*. Annals of Botany. 2009;103: 403-410, doi: 10.1093/aob/mcn229

[15] Srivastava AK, Sablok G, Hackenberg M, Deshpande U,

Suprasanna P. Thiourea priming enhances salt tolerance through co-ordinated regulation of microRNAs and hormones in Brassica juncea. Scientific Reports. 2017;7, 45490. doi: 10.1038/srep45490.

[16] Isayenkov S, Maathuis FJM. Plant salinity stress; many unanswered questions remain. Frontiers in Plant Science. 2019;10:1-11. 10.3389/fpls.2019.00080

[17] Rodriguez RJ, Henson J, Van Volkenburgh E, Hoy M, Wright L, Beckwith F, Kim YO, Redman RS. Stress tolerance in plants via habitat-adapted symbiosis. The ISME Journal. 2008;2:404-416.

[18] Shekari F, Soltaniband V, Javanmard A, Abbasi A. The impact of drought stress at different stages of development on water relations, stomatal density and quality changes of rapeseed(*Brassica napus* L.). Iran Agricultural Research. 2015;34(2):81-90.

[19] Hasanuzzaman M, Nahar K, Hossain MS, Mahmud JA, Rahman A, Inafuku M, Fujita M. Coordinated actions of glyoxalase and antioxidant defense systems in conferring abiotic stress tolerance in plants. International Journal of Molecular Sciences. 2017;18(1):200-228.

[20] Kaur S, Sharma P. Biochemical characterization of Indian mustard (*Brassica juncea* L.) genotypes in response to moisture stress and irrigation modules. Journal of Oilseed Brassica. 2015;6(2):265-272.

[21] Cheema, M.A.;Wahid, M.A.; Sattar, A.; Rasul, F.; Saleem, M.F. Influence of different levels of potassium on growth, yield and quality of canola (*Brassica napus* L.) cultivars. Pakistan Journal of Agriculture Science. 2012; 49: 163-168.

[22] Yamaguchi-Shinozaki K, Shinozaki K. Transcriptional regulatory networks in cellular responses and tolerance Of dehydration and cold stresses. Annual Reviews in Plant Biology. 2006;57:781-803.

[23] Sakuma Y, Maruyama K, Osakabe Y, Qin F, Seki M, Shinozaki K. Functional analysis of an Arabidopsis transcription factor, DREB2A, involved in drought-responsive gene expression. Plant Cell. 2006;18:1292-1309.

[24] Dreccer MF, Fainges J, Whish J, Ogbonnaya FC, Sadras VO. Comparison of sensitive stages of wheat, barley, canola, chickpea and field pea to temperature and water stress across Australia. Agricultural Meteorology. 2018;248: 275-294. 10.1016/j.agrformet.2017.10.006

[25] Shah SH, Ali S, Hussain Z, Jan SA, Din JU, Ali GM. Genetic improvement of tomato (*Solanum lycopersicum*) with AtDREB1A gene for cold stress tolerance using optimized Agrobacterium-mediated transformation system. International Journal of Agriculture and Biology. 2016;18:471-482.

[26] Patade VY, Khatri D, Manoj K, Kumari M, Ahmed Z. Cold tolerance in thiourea primed capsicum seedlings is associated with transcript regulation of stress responsive genes. Molecular Biology Reports. 2012;39:10603-10613. 10.1007/s11033-012-1948-6.

[27] Proshad R, Kormoker T, Mursheed N, Islam MM, Bhuyan MI, Islam MS, Mithu TN. Heavy metal toxicity in agricultural soil due to rapid industrialization in Bangladesh: a review. International Journal of Advanced Geosciences. 2018;6 (1), 83-88.

[28] Minhas P, Rane J, Pasala RK (Eds.). Abiotic Stress Management for Resilient Agriculture. (Singapore: Springer;), ISBN 978-981-10-5744-1.

[29] Khan ZI, Ugulu I, Sahira S, Ahmad K, Ashfaq A, Mehmood N.

Determination of toxic metals in fruits of *Abelmoschus esculentus* grown in contaminated soils with different irrigation sources by spectroscopic method. International Journal of Environmental Research. 2018;12:503-511. 10.1007/s41742-018-0110-2.

[30] Narendrula-Kotha R, Theriault G, Mehes-Smith M, Kalubi K, Nkongolo K. Metal toxicity and resistance in plants and microorganisms in terrestrial ecosystems. Reviews of Environmental Contamination and Toxicology. 2020;249, 1-27. 10.1007/398_2018_22.

[31] Khalid S, Shahid M, Bibi I, Sarwar T, Shah A, Niazi N. A review of environmental contamination and health risk assessment of wastewater use for crop irrigation with a focus on low and high-income countries. International Journal of Environmental Research. 2018;15:895. 10.3390/ijerph15050895.

[32] Couto R, Comin JJ, Souza M, Ricachenevsky FK, Lana M, Gatiboni L. Should heavy metals be monitored in foods derived from soils fertilized with animal waste? Frontiers in Plant Science. 2018;9: 732. 10.3389/fpls.2018.00732.

[33] Wang JL, Li T, Liu GY, Smith JM, Zhao ZW. Unravelling the role of dark septate endophyte (DSE) colonizing maize (*Zea mays*) under cadmium stress: physiological, cytological and genic aspects. Scientific Reports. 2016; 6: 22028

[34] Friesen ML, Porter SS, Stark SC, Von Wettberg EJ, Sachs JL, Martinez-Romero E. Microbially mediated plant functional traits. Annual Review of Ecology, Evolution, and Systematics. 2011;42: 23-46.

[35] Hu Z, Zhang D, Zhang G, Kan G, Hong D, Yu D. Association mapping of yield-related traits and SSR markers in wild soybean (*Glycine soja* Sieb. and Zucc.). Breeding Science. 2014;63:441-449

[36] Qu C, Jia L, Fu F, Zhao H. Genome-wide association mapping and Identification of candidate genes for fatty acid composition in *Brassica napus* L. using SNP markers. BMC Genomics. 2017;18:232.

[37] Zhu Q, King GJ, Liu X, Shan N, Borpatragohain P, Baten A, Wang P, Luo S, Zhou Q. Identification of SNP loci and candidate genes related to four important fatty acid composition in *Brassica napus* using genome wide association study. PLoS One. 2019;14:e0221578.

[38] Lu K, Wei L, Li X, Wang Y, Wu J. Whole-genome resequencing reveals *Brassica napus* origin and genetic loci involved in its improvement. Nature Communications. 2019;10:1154

[39] Saharan GS, Mehta N and Sangwan MS. Diseases of oilseed crops. Indus Publication Co., New Delhi. 2005. ISBN 10: 8173871760 / ISBN 13: 9788173871764

[40] Saha LR and Singh HB. Diseases of rapeseed and mustard and their management. Review of Tropical Plant Pathology. 1988;5:47-77

[41] Kolte SJ. Tackling fungal diseases of oil seed Brassicas in India. Brassica. 2005;7 (1/2): 7-13.

[42] Mehta N, Sangwan MS, Saharan GS. Fungal diseases of rapeseed mustard. Diseases of Oil Seed Crops. 2005;15-86.

[43] Kotle SJ. Diseases of Annual Edible Oilseed Crop. Rapeseed mustard and sesame diseases. Vol.II. C.R.C. Press, Boca Raton, Florida, U.S.A. 1985;320.I SBN: 9781351071437

[44] Singh RK Prasad A, Muthamilarasan M, Parida SK, Prasad M. Breeding and

biotechnological interventions for trait improvement: status and prospects. Planta. 2020;252 (54):3-18. https://doi.org/10.1007/s00425-020-03465-4

[45] Manoj K, Rai MK, Kalia RK, Rohtas RS, Dhawan A. Developing stress tolerant plants through in vitro selection—An overview of the recent progress. Environmental and Experimental Botany. 71(1):89-98. DOI: 10.1016/j.envexpbot.2010.10.021

[46] Rai MK, Rajwant KK, Rohtas S, Manu PG, Dhawan AK. Developing stress tolerant plants through in vitro selection – an overview of the recent progress. Environmental and Experimental Botany. 2011;71:89-98.

[47] Pérez-Clemente RM, Gómez-Cadenas A. 2012; In vitro tissue culture, a tool for the study and breeding of plants subjected to abiotic stress conditions. Book chapter In Techopen. 2012 DOI: 10.5772/50671

[48] MacDonald MV, Ingram DS. In vitro selection for resistance to *Alternaria brassicicola* in *Brassica napus* spp. oleifera (winter oilseed rape) using partially purified culture filtrates. Cruciferae Newsletter. 1985;10:97-100.

[49] Sacristan MD. Resistance response to *Phoma lingum* of plant regenerated from selected cells and embryogenic cultures of haploid Brassica napus. The or Appl Genet. 1982;61:193-200.

[50] Ganesan M, Jayabalan N. Isolation of disease-tolerant cotton (*Gossypium hirsutum* L. cv. SVPR 2) plants by screening somatic embryos with fungal culture filtrate. Plant Cell, Tissue and Organ Culture. 2006; 87:273-284.

[51] Kumar JV, Ranjitha Kumari BD, Sujatha G, Castano E. Production of plants resistant to *Alternaria carthami* via organogenesis and somatic embryogenesis of safflower cv. NARI-6 treated with fungal culture filtrates. Plant Cell, Tissue and Organ Culture. 2008; 93:85-96.

[52] Jayashankar S, Li Z, Gray DJ. In vitro selection of *Vitis vinifera* Chardonnay with Elsinoe ampelina culture filtrate is accompanied by fungal resistance and enhanced secretion of chitinase. Planta 2000;211:200-208

[53] Lewis DH. Symbiosis and mutualism: crisp concepts and soggy semantics. The Biology of Mutualism: Ecology and Evolution, ed. Boucher. 1985;29-39.

[54] Gill SS, Tuteja N. Reactive oxygen species and antioxidant machinery in abiotic stress tolerance in crop plants. Plant Physiology and Biochemistry. 2010;48: 909-930.

[55] Pieterse CM, Zamioudis C, Berendsen RL, Weller DM, Van Wees SC, Bakker PA. Induced systemic resistance by beneficial microbes. Annual Reviews in Phytopathology. 2014;52:347-375.

[56] Choudhary DK, Johri BN. Interactions of *Bacillus* spp. and plants - with special reference to induced systemic resistance (ISR). Microbiological Research. 2009;164:493-513.

[57] Vishwanath, Kolte SJ. Methods of inoculation for resistance to Alternaria blight of rapeseed and mustard. Journal of Mycology and Plant Pathology. 1999; 29: 96-99.

[58] Ardanov P, Ovcharenko L, Zaets I, Kozyrovska N, Pirttila AM. Endophytic bacteria enhancing growth and disease resistance of potato (*Solanum tuberosum* L.). Biological Control. 2011;56:43-49

Chapter 4

Embryo Culture and Embryo Rescue in *Brassica*

Mohammad Akmal

Abstract

Somatic embryogenesis is the best demonstration of totipotency in higher plants in which somatic cell produce whole plant like zygotic embryo. It is also demonstrated that immature, weak, hybrid or sometimes inviable embryos can be saved through *in vitro* culture to prevents its degradation. It may help to cross the reproductive barriers when interspecific hybrids developed. *Brasssica* is an economically valuable oil yielding and vegetable crop and India is the largest producer of oil seed rape in the world. Various factors affect the embryo rescue in *Brassica* like growth stage of the embryos, types and composition of the rescue medium etc. The embryo regeneration potential can improve through the modification of culture conditions in both zygotic as well as somatic embryo. Except the embryo culture other parts like ovule, ovary culture can also be done to developed interspecific hybrids. This chapter is focused on the embryo rescue techniques in the genus *Brassica* and summarizes possible ways of improving the technique used.

Keywords: *Brassica*, embryo rescue, hybrids, embryo culture, somatic embryos

1. Introduction

Brassica is an important vegetable and oilseed crop of India and it is the largest producer of the oilseed rape in the world. The genus belongs to the family Brassicaceae having about 38 different species [1]. Crossing and hybridization in these species was done to developed new cultivars worldwide for improving traits and yields. The naturally occurring genetic variation is the basis of the improvements in *Brassica* to produce new morphotypes in interspecific hybridization program [2]. The interspecific hybridization is a difficult process due to pre and post-fertilization barriers and abortion of the hybrid embryo. The embryo degeneration after some hybridization experiments takes place very early [3]. This may be due to the poor endosperm development or sometimes endosperm may not be developed [4]. But in non-endospermic embryos the post fertilization barrier cannot crossed and embryos are defective, disformed and aborted early. The hybrid embryos can be regenerated through various technique that's comes under the embryo rescue. The embryo rescue by culturing on nutrient medium that would support and orderly development of embryos. If the embryos are disformed, secondary embryogenesis may be induced by the manipulation of medium and growth regulators combination [5]. New biotechnological tool like transgenic technology would be better in improving the varieties of *Brassica* crop. Large number of transgenics, both biotic and abiotic stress resistant plants were developed [6, 7]. It is desired to produce transformed plant through somatic embryogenesis. It is preferred because of the genetic stability during culture and very low rate of genetic variations which is otherwise results into somatic variations or somaclones [8]. There are however, some exceptions as in wheat (*Triticum durum* Desf.) somaclonal variation had been reported for several agronomic and phenotypic traits, such as plant height, leaf size, pollen fertility, tolerance to aluminum toxicity, albinism and leaf malformations [9]. *Cymbopogon winterianus* Jowitt showed somaclonal variations when regenerated via somatic embryogenesis [10]. But these variations are restricted to only few plant genera and it is found that the somatic variation produced frequently during tissue culture but generally not in somatic embryo or secondary embryos culture.

The embryogenesis in higher plants is controlled by many genes and related embryos specific proteins. Loss of function mutants were used to identified the genes involved in There are about 220 *EMB* genes in *Arabidopsis* required for normal embryo development. These were identified through duplication of alleles or molecular complementation [11]. Some proteins also required for the somatic embryogenesis and in embryo rescue in Carrot (*Dacus carrota*), these are mainly glycoproteins secreted into the culture medium such as endichitinase and arabinigalactan proteins (AGPs) are required for somatic embryogenesis. These AGPs contains glucosamine and *N*-acetyl-*D*-glucosaminyl [12]. *Rhizobium leguminosarum nod* gene metabolic product contains *N*-acetylglucosamine lipooligosaccharides that promotes carrot embryo rescue [13].

2. Zygotic embryos, genetic embryos and somatic embryos

In higher plants, formation and development of embryo are the two distinct phenomena that can takes place inside the ovule. An ovule has different parts like chalaza, nucellus, micropyle, integuments, and most important embryo sac. The embryo sac develops after the reduction division of the megaspore mother cell. Only one haploid megaspore cell give rise to the birth of embryo sac that's contains many cells like synergid cells, antipodal cells and one egg cell. Instead of all these cells there is a central nucleus (So called because it is not surrounded by distinct cell wall). The egg cell is polar and contain a distinct nucleus on cytoplasm-rich chalazal pole while there are vacuoles at micropylar end [14]. Egg cell when fertilized called the zygote and gives rise to the embryo. The two important process alternate in each generation, the meiosis (reduction division resulting two haploid cells called the gametophyte) and the fertilization (fusion of the two-haploid nucleus called the sporophyte). The three process are distinct with one another. Formation of zygote after fertilization, formation of the embryo and development of the embryo. Failure of any process disturbed the embryogenesis and cease the formation and development of embryo. After the formation of complete embryo successfully the mature embryo enters in desiccated and metabolically quiescent state [15] or it undergo the period of dormancy that's complete the process. The dormancy is the process of adaptation to withstand unfavorable conditions. The zygotic embryo in the angiosperm after maturation develops into seeds and it is composed of several tissues, including the embryo, the endosperm, and the testa. The mature embryo has the bipolar axis, on which root and shoot meristem are present and gives rise to the root and shoot during plant development. In contrast to the zygotic embryo, the genetic embryo is formed without fertilization through the process termed as apomixis. Apomixis refers to the formation of embryo in the ovule from the somatic cells. These somatic cells when diploid i.e. nucellus or integument directly gives rise to the embryo and termed as sporophytic apomixis. However, when embryo sac originates either from megaspore mother cells by mitosis or incomplete meiosis in diplospory is termed as gametophytic apomixis.

Totipotency is the most spectacular demonstration of potencies in the cells of higher plants. The somatic embryogenesis is the generation of bipolar structure from any somatic cell that have distinct root and shoot pole. It is very rare phenomenon present in nature and restricted to some plants like *Kalenchoe*, *Bryophyllum* etc. But it may be introduced into other plants artificially through tissue culture technique. How somatic cell triggered to gives rise to the embryo? It is somewhat unclear but it is suggested that the irregular distribution of auxin stimulates the establishment of embryonic structure [16]. However, the stages of the development of the somatic embryos resemble with the zygotic embryos.

3. Embryo culture

In *Brassica*, somatic embryogenesis can be induced using various auxins like 2,4-D and NAA at higher concentrations separately or in combinations with the cytokinin [17]. 2,4-D alone proved best hormone that used to induced polarity in the somatic cells as compared to the combination with the cytokinin like kinetin as it may ceased the further proliferation of somatic embryos [18]. Various culture medium was used with the auxins like SH Medium [19], B5 medium [20], MS medium [21], Kao's medium [22] etc., but rapid propagation of somatic embryos in *Brassica* proved to be the best in MS medium [17]. MS basal medium and low pH (3.5–5) was also used to induced somatic embryos in *Brassica napus* using immature seeds 14 to 28 days after pollination [23, 24]. Not only in *B. napus* but in *B. oleracea* varieties, cabbage and cauliflower immature zygotic embryos gives rise to the somatic embryos with high frequency [5], confirming that the stress condition either due to the PGR (mainly auxin) reprogrammed the zygotic immature cell to induced embryogenesis in *Brassica*. There is not only induction and establishment of polarity but also maintenance of the root and shoot meristem. It is also noticed that the zygotic embryos when immature have higher embryogenic potential than the mature embryo [23, 25]. The low pH value i.e., 3.5–5 in *Brassica napus* increase the exchange of ions and ionic nutrients that's accumulate inside the embryogenic tissue or callus. The stored food material mainly lipid bodies in the cells of *Brassica* indicates the good exchange of nutrient under auxin induced stress condition. The leaf explant is best for the induction of somatic embryos as compared to the other explants like stem and hypocotyl sections (**Figure 1A-D** and **F**) [17, 26, 27]. This is because the large number of vacuolated protoplasts that's provide the space for the storage of the food material in cotyledon and leaf explants.

4. Embryo rescue in *Brassica*

Embryo rescue is an *in vitro*-culture technique that is used to save weak, immature and hybrid or sometimes inviable embryos to prevents its degradation. The procedure involves excising weak, immature plant embryos and culture them on specially devised culture medium. It plays an important role in plant breeding of important crop plants. In *Brassica*, the interspecific and intergeneric hybridization was done from very earlier because the crop yield losses due to disease, biotic and abiotic stresses is very high. Interspecific, intergeneric and intervarietal hybrids have been generated in mango, banana, seedless grape, papaya and seedless citrus using embryo rescue [28]. It is not only an important oilseed crop but used as vegetable, and fodder for the farm animals. The earlier attempts were made in Chinese cabbage varieties i.e. *B. oleracea* and *B. campestris* [3]. The hybrid embryos were carefully removed and cultured on the nutrient medium *in vitro*.

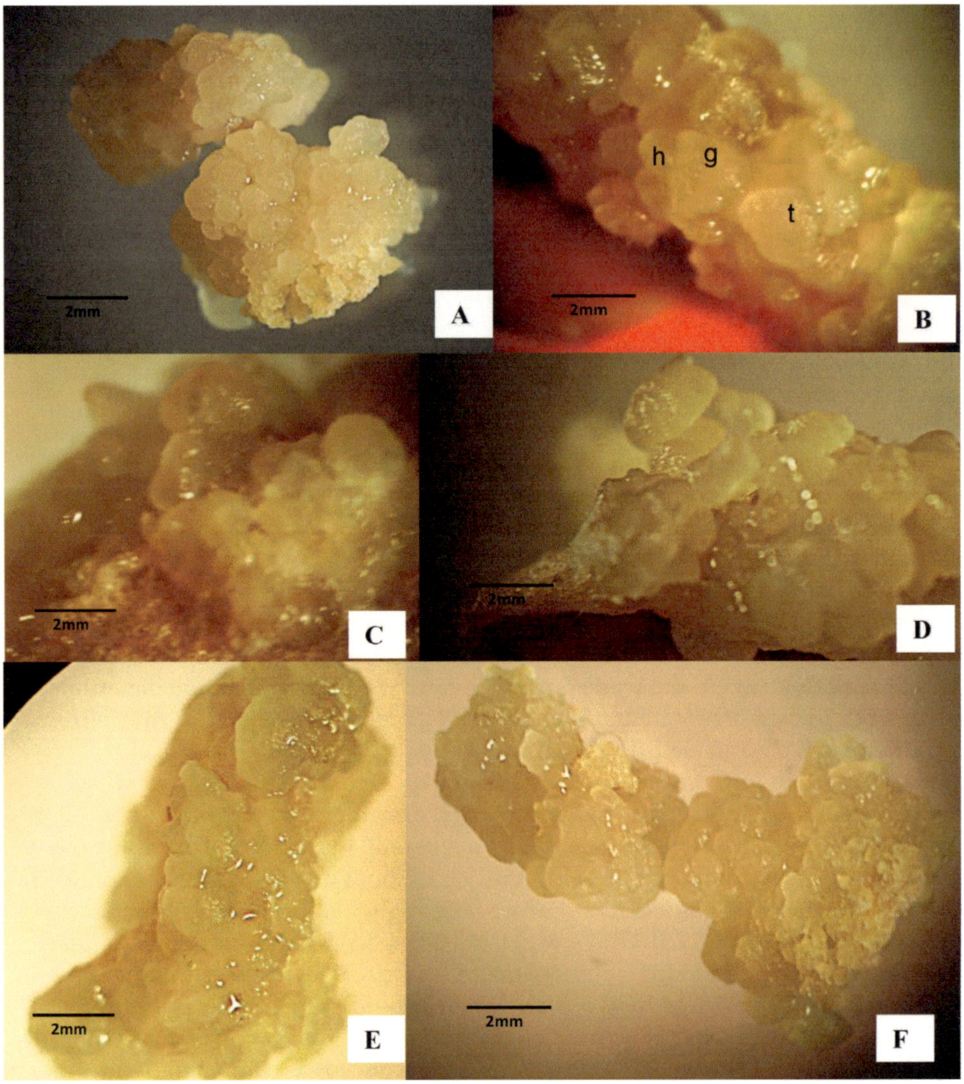

Figure 1.
Somatic embryos of Brassica juncea *L. A, B, C, and D, The cotyledon derived somatic Embryos, E and F, hypocotyl derived somatic embryos.*

The survival rate of the embryos can be increased when rescued because wide hybridization crosses fail to complete normal sexual reproduction cycle. In *Brassica* the best adopted methods for the embryo rescue are the direct ovule culture, siliqua culture and immature embryos culture. Sometimes embryo-nurse endosperm for embryo transplant was also adopted [29]. Very young embryo is difficult to culture on artificial culture medium. Due to this young fertilized pistil were cultured. Therefore, the chances of embryo abortion can be minimized. For siliqua culture young fertilized pistils excised 4 to 6 days after pollination and cultured on the MS medium. It absorbed the medium start to grow gradually but regular subculturing is required. The swollen pistil again excised to dissect out developing ovule. Ovules can also be culture in a similar manner and selection day after pollination may varies from plant to plant and to identify the fertilized pistil, pollinated pistil fixed in 70% ethanol 24 to 48 hour after pollination and stained in aniline blue, the pollen germination and pollen tube growth may be observed under the microscope [30]. Ovary culture in *Brassica* can be done 4 to 14 days after pollination. The stalk of the ovary cut from the base and cultured on the nutrient medium [31]. For embryos culture, the siliqua was collected after 10, 15, 20, 25, and 30 days after pollination and after the sterilization the ovules are dissect out from siliquae and the young embryos cultured on the MS medium [21]. All these methods of embryo rescue depend on the type and the condition of the hybrid embryos and the type of hybridization experiments [32]. Considerable progress in embryo rescue has been takes place but the rescue of hybrid embryos seems to be difficult when abortion occurs at very early. Embryo implantation technique was adopted to overcome this problem which is known as embryo-nurse endosperm transplant where excised hybrid embryo inserted into a cellular endosperm dissected from third species or one of the parents. The nurse endosperm with the transplanted embryo cultured on an artificial medium [33]. Recently double haploids were developed through embryoids derived from isolated microspore culture [34].

5. Factors affecting embryo rescue

Various factors affect the embryo rescue in *Brassica* and other plants. It depends on the age of the embryo, intactness of the suspensor [35], excision procedure, sterilization, culture medium supplementation, temperature and light requirements etc. Highly immature embryo rescue is very difficult and requires special medium requirements. Excision procedure is designed in that way, no or minimum injury should be there in embryo proper specially when embryo excised at very early stage of development. The aim should be rescue for primary embryos secondary embryogenesis should be avoided.

5.1 Nurse tissue

The early stage embryo culture during rescue is very difficult to culture on artificial medium. The nurse tissue provides a natural condition and nutrition so that the chances of embryo abortion reduces. The best nurse tissue for the embryo culture is endosperm and if it is some days old it efficiently increases the chances of survival of the young embryo. The somatic embryos can be used as nurse tissue for culturing the hybrid zygotic embryos as done in *Pinus* [36]. The ovule is another nurse tissue which is used to generate very young stage embryos. It is not necessary to excise young embryo from the ovule but excision of ovule is easy. Sometime ovary culture can also be used and this was done when the embryos are very small and inconspicuous.

5.2 Culture medium

There are various nutrients formulations used to culture the embryos during rescue. Murashige and Skoog formulation [21] and Gamborg's B-5 [20] are most frequently used in *Brassica* [37]. Sometimes both hormones were used as in *B. oleracea* [5]. However, there are examples in which B5 vitamins are used with MS salts in embryo rescue of *R. sativus* [38]. Other mediums used for the embryo rescue in other plants are Knop's medium; Heller's medium [39], Monnier's medium [40] etc. The early heterotrophic immature embryos take its nutrition form the endosperms and surrounding tissue but when embryos mature it is partly autotrophic and it requires only basic mineral salts and sucrose. In *Brassica*, the embryos become autotrophic very late after globular stages [41]. The early globular stage of proembryo culture was achieved earlier using double-layer culture system and in embryo culture medium which contains mineral salts, sugars, amino acids, organic acids and coconut water [42]. The coconut water induces cell division in plant and necessary for the development of very young embryo [43]. The other additive components that used during embryo culture are tomato juice, banana pulp, different fruit juices, fish emulsion, leaf extract, potato extract etc. [44]. Sucrose is a very important constituent of the culture medium as an energy source and it maintains the osmotic potential of culture medium. High sucrose contents trigger the formation of embryos because it mimics the high osmotic potential of the embryo sac [29], and it help for the induction of embryos from embryogenic calli [17].

5.3 Silver nitrate

The silver nitrate is another very important ethylene antagonistic component that's play important role in *Brassica* embryo culture. It is added into the medium at a concentration that varies between (1–10 mgl^{-1}). It significantly increases the regeneration potential in various *Brassica* species through both embryogenesis and organogenesis. During somatic embryogenesis the presence of $AgNO_3$ increase the no of embryos significantly specially in *B. oleracea* and *B. rapa* [45].

5.4 Temperature and light

Embryo culture during embryo rescue influenced with temperature and light and their requirements are also varies in different plants [46]. The low temperature regime is best for the embryo rescue but in some cases at high temperature regime (26.4 °C/10.4 °C embryo rescue through ovule culture method gives better results than low temperature regime period (19.4 °C/4.3 °C) (1.75%) in *B. oleracea* [47]. According to Mei et al., [48] there is a significant quadric relation between the effective accumulated temperature (EAT) and the efficacy of ovule culture. The hybrid *B. napus* x *B. oleracea* siliqua can be collect for the excision of ovule on the basis of EAT instead of siliqua age.

6. Cytology of cultured embryos

Several biochemical and molecular changes occurs at cellular level as the embryo formed, grow in length and approaches to the maturity. The developmental studies of the somatic embryos showed that the lipid, protein and polysaccharides produced during at varying degrees when embryos were cultured on ABA containing medium. At first one or two weeks polysaccharides were produced and after that polysaccharides lipid and protein accumulated [49]. The analysis of the cultured embryos cells of oil yielding crop like *Brassica*, under electron microscopy revealed the presence of lipid bodies (spherosomes) associated with the endoplasmic reticulum [50]. The embryos cell contains several other components like some dense granules (ribonucleoprotein), endoplasmic reticulum, mitochondria, amyloplast and some irregular bodies etc. It was noted earlier that when a cell start to convert into an embryo, it start to contains smaller vacuoles in dance cytoplasm, large nucleus (Nu) with numerous organelles and stored bodies [17]. These are the steps towards the development of seed and the stored food material is the reserve for germination. The stored food material in the small somatic embryos is similar to as in the endosperms of zygotic embryos as in *Acromia aculeata*. The endosperm of zygotic embryos showed the accumulation of lipid and proteins which may consumed in the initial stages of germination and plantlet establishment (**Figure 2A-E**). However, their somatic embryos does not showed such types of deposition and this results in low conversion of these embryos into plants [51]. In other plants like orchids some early stage protocorm behave like somatic embryos and contains protein bodies and starch granules and it is supposed to be similar with the zygotic embryo development [52]. The deposition of the polysaccharides starts toward the basal end of the suspensor

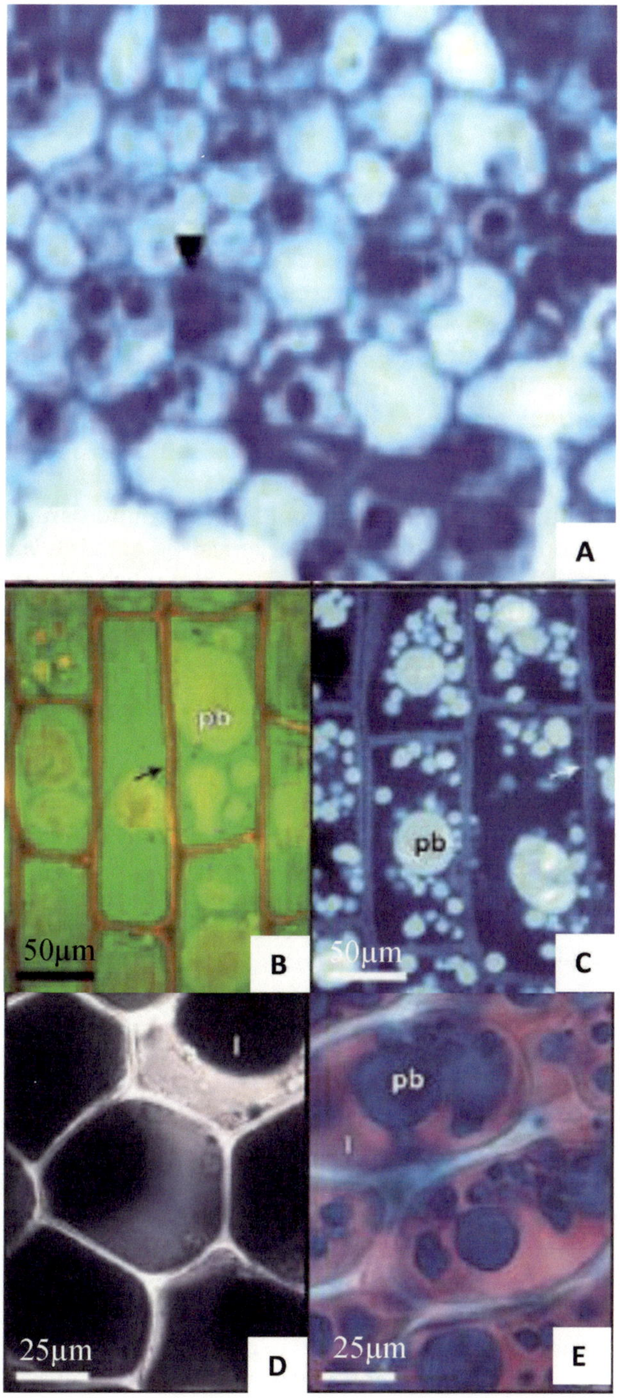

Figure 2.
A Light micrographs of Brassica juncea *somatic embryo sections, cells showed the stored food material in the form of lipid B, C, D and E Light micrographs of different histochemical tests of* Acromia aculeata *endosperm of zygotic embryos showed lipid (l) and protein bodies (pb). Source: Akmal et al [17], Moura et al. [51].*

and gradually it increases and move inwards in the cortex of developing embryos. The stored lipid bodies in embryo at its peak during the cotyledon development and after that decreases. The most abundant fatty acids in somatic embryos of *Picea abies* are linolic, oleic, palmitic and 5,9-octadecanioc acids [53]. There is a quantitative difference in the composition of the fatty acids in *in vitro* and *in vivo* cultured somatic embryos [54].

7. Genes involved in embryogenesis

When a somatic cell enters into embryogenic state it starts to modify the gene expression level. There are number of genes involved in somatic embryogenesis, showed increase expression. These can be categorized in various groups [55]. These are hormone responsive genes, house-keeping genes, genes expressed during maturation, genes coding for extracellular proteins, homeotic genes, HSPs,(Heat-Shock Proteins) germins, zygotic mutants and genes for signal transductions, except these there are the genes of transcription factors. The various transcription factors regulate several genes that showed expression in the somatic embryo induction [15]. In *Arabidopsis thaliana* the basic pattern of embryo formation is the polarity of apical-basal axis and other perpendicular to the axis. It consists of shoot meristem, cotyledons, hypocotyl, root, and root meristem along the apical–basal axis and a concentric arrangement of epidermis, subepidermal ground tissue, and central vascular cylinder along the radial axis For the apical-basal axis formation asymmetric *PIN7* and *WOX2* expression is important to establish apical cell identity at the apical and basal cell of the zygote while another *gnom* mutant argues against the possibility that the different sizes of the daughter cells per se are important for apical–basal development [56]. There are some genes that interact with one another and control the pattern formation. The mutant analysis showed that the resulting mutant disturbed the pattern formation related to the primary shoot and root meristems [57]. Some embryogenesis-related genes similar expression pattern in both in vitro and in vivo embryogenesis and these were *LEA* (Late Embryogenesis Abundance) genes, *SERK*, (*SOMATIC EMBRYOGENESIS RECEPTOR-LIKE KINASE*), *AGL15 (AGAMOUS-LIKE15*), *BBM* (BABY BOOM gene*)*, *LEC1*(LEAFY COTYLEDON*)*, *FUS3* (FUSCA3, B3-domin transcription factor) and *AB13 (ABSCICSIC ACID INSENSETIVE3)* [58]. There are about 250 EMB genes required for the normal development of the embryo and for complete seed development additional 219 genes product required [11, 59]. A dataset of about 510 *EMBRYO-DEFECTIVE* (*EMB*) genes were identified in Arabidopsis [60]. The TARGET OF RAPAMYCIN (TOR) kinase has been recognize as a key developmental regulator in both plants and animals which integrates environmental and nutrient signals to direct growth and development [61]. The rapamycin kinase receptor gene in *Arabidopsis.*

(AtRaptor1) is responsible for the early development of embryo [62]. The genes also essential for the post-embryonic plant growth because the AtRaptor1A, AtRaptor1B double mutants are defective in meristem driven growth during post embryonic stage [63].

In near future, the techniques of somatic embryos induction, and embryo rescue in *Brassica* can be further improved so that there should be about 99% chances of hybrid embryos survival. This will not only increase their survival rate but number of plantlets regenerated also improved, especially in those intergeneric crosses where the hybrid embryos survival rate is very low. This will give an opportunity to increase the productivity of the *Brassica* crop through various crop improvement programs.

8. Conclusion

In this chapter, the somatic embryos and zygotic embryos with embryo rescue techniques are discussed. The hybrid embryo rescue technique in *Brassica* and other plants provide a useful tool to developed intergeneric and interspecific hybrids. The modification of the culture conditions, use of plant growth regulators and other complex medium components, the immature hybrid embryo can be successfully rescued at an early stage. The hybrid embryo in *Brassica* directly gives rise to new plants or some time it is desired to generate secondary embryos. The cytology of the somatic and zygotic embryos and the genes involved are also briefly discussed.

Acknowledgements

The author acknowledges Prof. M. Z. Abdin lab, Jamia Hamdard, New Delhi for carried out research work on somatic embryogenesis in *Brassica juncea* L.

Acronyms and abbreviations

LEA	Late Embryogenesis Abundance
BBM	BABY BOOM gene
SERK, LEC	Leafy cotyledon
AB13	Abscicsic acid insensetive3
AGL15	Agamous-like15
WUS	WUSCHEL
SERC	Somatic embryogenesis receptor-like kinase

Author details

Mohammad Akmal
Advanced Plant Physiology Laboratory, Department of Botany, M. L. K. PG. College, Balrampur, India

*Address all correspondence to: akmal729@gmail.com

References

[1] Cheng F, Mandáková T, Wu J, Xie Q, Lysak MA. Wang X. Deciphering the diploid ancestral genome of the mesohexaploid *Brassica rapa*. Plant Cell. 2013;25 (5):1541-1554. DOI: 10.1105/tpc.113.110486

[2] Pen S, Nath UK, Song S, et al. Developmental stage and shape of embryo determine the efficacy of embryo rescue in introgressing orange/yellow color and anthocyanin genes of *Brassica* species. Plants (Basel). 2018;7(4):99. DOI: 10.3390/plants7040099

[3] Agnihotri A. Hybrid embryo rescue. In: Lindsey K, ed. *Plant Tissue Culture Manual: Fundamentals and Applications*. Dordrecht, The Netherlands: Kluwer Academic Publishers, 1993; E4.1–E4.8.

[4] Shukla S. Embryo rescue technology: An approach for varietal development and *in vitro* germplasm conservation. National Academy of Agriccuture Science 2016;34(3):841-847.

[5] Pavlović S, Vinterhalter B, Zdravković-Korać S, et al. Recurrent somatic embryogenesis and plant regeneration from immature zygotic embryos of cabbage (*Brassica oleracea* var. *capitata*) and cauliflower (*Brassica oleracea* var. *botrytis*). Plant Cell, Tissue and Organ Culture 2012;113 397-406. DOI:10.1007/s11240-012-0279-6

[6] Shah N, Anwar S, Xu J, Salah A, et al. The response of transgenic *Brassica* species to salt stress: a review. Biotechnology Letters. 2018;40:1159-1165. DOI: 10.1007/s10529-018-2570-z

[7] Zhuang W, Tianyu L, Xiaochun S, Hongxue W, et al. Overexpression of MzASMT 1, a gene from malus zumi mats, enhances salt tolerance in transgenic tobacco. Frontiers in Plant Science 2020;11:1590. DOI: 10.3389/fpls.2020.561903

[8] Narvaez I, Martin C, Jiménez-Díaz RM, Mercado JA, Pliego-Alfaro F. Plant regeneration via somatic embryogenesis in mature wild olive genotypes resistant to the defoliating pathotype of *Verticillium dahlia*. Frontiers in Plant Science. 2019;10:1471. DOI: 10.3389/fpls.2019.01471

[9] Bouiamrine, E.H., M. Diouri and R. El-Halimi,. Assessment of somaclonal variation in regenerated plants from immature embryos culture of durum wheat. International Journal of Agricuture and Biology. 2012;14: 941-946

[10] Dey T, Saha S, Ghosh PD. Somaclonal variation among somatic embryo derived plants — Evaluation of agronomically important somaclones and detection of genetic changes by RAPD in Cymbopogon winterianus, South African Journal of Botany. 2015;96:112-121. DOI: 10.1016/j.sajb.2014.10.010.

[11] Tzafrir I, Pena-Muralla R, Dickerman A et al. Identification of genes required for embryo development in *Arabidopsis*. Plant Physiology. 2004;135(3):1206-1220. DOI: 10.1104/pp.104.045179

[12] van Hengel AJ, Tadesse Z, Immerzeel P, Schols H, van Kammen A, de Vries SC. N-acetylglucosamine and glucosamine-containing arabinogalactan proteins control somatic embryogenesis. Plant Physiology. 2001;125(4):1880-1890. doi:10.1104/pp.125.4.1880

[13] De Jong AJ, Heidstra R, Spaink HP, Hartog MV et al. Rhizobium lipo-oligoscharides rescue a *Daucus carota* somatic embryo variant. Plant cell.1993;5:615-620.

[14] Dodeman VL, Ducreux G, Kresis M. Zygotic embryogenesis versus somatic

embryogenesis. Journal of Experimental Botany 1997;48(313):1493-1509.

[15] Méndez-Hernández Hugo A., Ledezma-Rodríguez Maharshi, Avilez-Montalvo Randy N. et al. Signaling Overview of Plant Somatic Embryogenesis. Frontiers in Plant Science. 2019;10:77. DOI: 10.3389/fpls.2019.00077

[16] Márquez-López RE, Pérez-Hernández, CA, Kú-González Á, Galaz-Ávalos RM, Loyola-Vargas, VM. Localization and transport of indole- 3-acetic acid during somatic embryogenesis in *Coffea canephora*. Protoplasma. (2018);255:695-708. DOI: 10.1007/s00709-017-1181-1

[17] Akmal M, Nafis T, Mirza KJ et al. High frequency somatic embryogenesis in mustard crop (*Brassica juncea* L. cv. Pusa Jai kisan): Microscopic and histological analyses. Australian Journal of Crop Science 2011;5(13):1783-1789.

[18] Siong PK, Taha RM, Rehman FA. Somatic embryogenesis and plant regeneration from hypocotyl and leaf explants of Brassica oleracea var. botrytis (Califlower). Acta Biologica cracoviensia series Botanica. 2011;53(1):26-31. DOI: 10.2478/v10182-011-0004-5

[19] Schenk RU, Hildebrandt AC. Medium and technique for induction and growth of monocotyledonus and dicotyledonous plant cell culture. Canadian Journal of Botany. 2011;50(1):199-204.

[20] Gamborg OL, Murashige T, Thorpe TA, et al. Plant tissue culture media. *In Vitro* Cellular & Developmental Biology - Plant. 1976;12:473-478. DOI:10.1007/BF02796489

[21] Murashige T, Skoog F.. A revised medium for rapid growth and bioassays with tobacco tissue cultures. Physiologia Plantarum1962;15:473-497.

[22] Kao KN, Michayluk M. Nutritional requirements for growth of Vicia hajstana cells and protoplasts at a very low population density in liquid media. Planta. 1975;126:105-110.

[23] Koh WL, Loh CS. Direct somatic embryogenesis, plant regeneration and in vitro flowering in rapid -cycling *Brassica napus*. Plant Cell Report 2000;19:1177-1183.

[24] Burbuils N, Kupriene R, Liakas V. Somatic embryogenesis and plant regeneration in immature zygotic embryos of *Brassica napus*. Acta Universitatis Latviensis (Biology). 2007;723:27-35.

[25] Burbulis N, Kupriene R Induction of somatic embryos on *in vitro* cultured zygotic embryos of spring *Brassica napus*. Acta Universitatis Latviensis (Biology). 2005;691:37-143

[26] Narasimhulu SB, Kirti PB, Prakash S, Chopra VL. Somatic embryogenesis in Brassica nigra (Koch), Journal of Experimental Botany. 1992;43(9):1203-1207. DOI: 10.1093/jxb/43.9.1203

[27] Ripa RS, Arif MR, Islam MT, et al. Embryo rescue response and genetic analysis in interspecific crosses of oilseed *Brassica* species. In Vitro cellular & Developmental Biology - plant. 2020;56:682-693. DOI:10.1007/s11627-020-10116-6

[28] Sahijram L, Soneji JR, Naren A, Rao BM. Hybrid embryo rescue: a non-conventional breeding strategy in horticulture crops. Journal of Horticultural Sciences. 2013;8(1):1-20.

[29] Kumari P, Thaneshwari, Rahul. Embryo rescue in horticulture crops. International Journal of Current Microbiology and Applied Sciiences.

2018;7(6):3350-3358. DOI: 10.20456/ijcmas.2018.706.393

[30] Gupta D, Sharma SK. Embryo-ovule rescue technique for overcoming post fertilization barriers in interspecific crosses of *lens*. Journal of Lentil Research. 2005;2:27-30.

[31] Wen J, Tu Jx, Li Zy, et al. Improving ovary and embryo culture techniques for efficient resynthesis of *Brassica napus* from reciprocal crosses between yellow-seeded diploids *B. rapa* and *B. oleracea*. Euphytica. 2008;162:81-89. DOI: 10.007/s10681-007-9566-4

[32] Inomata N. Embryo Rescue Techniques for Wide Hybridization. In: Labana K.S., Banga S.S., Banga S.K. (eds) Breeding Oilseed *Brassicas*. Monographs on Theoretical and Applied Genetics, vol 19. Springer, Berlin, Heidelberg. 1993. P. 94-107. DOI: 10.1007/978-3-662-06166-4_7

[33] Bhojwani SS, Razdan MK, Zygotic embryo culture. In: Studies in Plant Science. 8th ed. Elsevier. 2008;5:297-335 DOI: 10.1016/S0928-3420(96)80013-9.

[34] Daurova A, Daurov D, Volkov D, Zhapar K, et al. Doubled haploids of interspecific hybrids between *Brassica napus* and *Brassica rapa* for canola production with valuable breeding traits. Oilseeds & fats Crops and Lipids. 2020;27:45. DOI: 10.1051/0cl/2020041

[35] Hu CY, Zanettini MHB. Embryo Culture and Embryo Rescue for Wide Cross Hybrids. In: Gamborg O.L., Phillips G.C. (eds) Plant Cell, Tissue and Organ Culture. Springer Lab Manual. Springer, Berlin, Heidelberg. 1995. P. 129-141. DOI: 10.1007/978-3-642-79048-5_11

[36] Hargreaves C, Reeves C, Gough K et al. Nurse Tissue for embryo rescue: testing new conifer somatic embryogenesis methods in a F1 hybrid pine. Trees. 2017;31:273-283.DOI: 10.1007/s00468-016-1482-6

[37] Burbulis N, Kupriene R, Zilenaite L. Embryogenesis, callogenesis and plant regeneration from anther culture of spring rape (*Brassica juncea* L.). Acta Universitatis. Latviensis, Biol. 2004; 676:153-158

[38] Mithila J, Hall JC. Transfer of auxinic herbicide resistance from wild mustard (*Sinapis arvensis*) into radish (*Raphanus sativus*) through embryo rescue. Journal of Horticultural Sciences. 2012;7(1):29-33.

[39] Carew DP, AE Schwarting. Production of Rye Embryo Callus. Botanical Gazette.1958;119(4):237-239.

[40] Monnier M. Culture of zygotic embryos. In: Thorpe TA, editors. *In Vitro* Embryogenesis in Plants. Current Plant Science and Biotechnology in Agriculture. Springer, Dordrecht; 1995. 20. DOI:10.007/978-94-011-0485-2_4

[41] Raghavan V. Experimental embryogenesis in vascular plants. Academic Press, 1976.London.

[42] Liu CM, Xu ZH, Chua NH. Proembryo culture: In vitro development of early globular-stage zygotic embryos from *Brassica juncea*. The Plant journal. 2011;3(2):291-300. DOI: 10.1111/j.1365-313X.1993.tb00179.x

[43] Thorpe T, Stasolla C, Yeung EC, de Klerk GJ, Roberts A, George EF. The components of plant tissue culture media II: Organic additions, osmotic and pH effects, and support systems. In Goerge EF, Hall MA, De Klerk GJ (eds), Plant Propagation by Tissue Culture. 2008; 1. The Background. 3rd ed. Springer, Dordrecht: 115-173.

[44] Reddy J. Nutrient media used for micropropagation of orchids: A research review. World Journal of Pharmaceutical

Research. 2019;5(9):1719-1732. DOI: 10.20959/wjpr20169-7036

[45] Kabir K, Kwon S-W, Park Y-J. Application of cobalt chloride and silver nitrate for efficient microspore culture of *Brassica rapa ssp. plant* Tissue Culture and Biotechnology. 2013;23(1), 1-10. DOI: 10.3329/ptcb.v23i1.15554

[46] Reed SM. Embryo rescue. Plant Development and Biotechnology. 2005; 235-239

[47] Sharma BB, Kalia P, Singh D and Sharma TR Introgression of black rot resistance from *Brassica carinata* to cauliflower (*Brassica oleracea* botrytis group) through embryo rescue. Frontiers in Plant Science. 2017;8:1255. DOI: 10.3389/fpls.2017.01255

[48] Mei JQ, DaYong W, Qinfei Li, et al Effective accumulated temperature is associated with the efficiency of hybrid ovary culture between *B. napus* and *B. oleracea*. Acta Physiologiae Plantarum. 2015;37(2):18. DOI: 10.1007/s11738-014-1763-x

[49] Joy RW, Young EC, Kong L. et al. Development of white spruce somatic embryos.: I. Storage product deposition. *In Vitro* Cellular & Developmental Biology - Plant. 1991;27:32-41. DOI: 10.1007/BF02632059

[50] Wann Wanner G, Formanek H, Theimer RR. The ontogeny of lipid bodies (spherosomes) in plants cells. Planta. 1981;151:109-123. DOI: 10.1007/BF00387812

[51] Moura EF, Ventrella MC, Motoike SY. Anatomy, histochemistry and ultrastructure of seed and somatic embryo of *Acrocomia aculeata* (Arecaceae) Scientia Agricola. (Piracicaba, Braz.). 2010;67(4):399-407.

[52] Lee Y-I, Hsu S-T, Yeung EC. Orchids protocorm-like bodies are somatic embryos. American Journal of Botany. 2013;100(11):2121-2131. DOI: 10.3732/ajb.1300193

[53] Grigová M, Kubes M, Drázná N, Rezanka T, Lipavská H. Storage lipid dynamics in somatic embryos of Norway spruce (Picea abies): histochemical and quantitative analyses. Tree Physiology 2007;27(11):1533-1540. DOI: 10.1093/treephys/27.11.1533

[54] Avjioglu A, Knox RB. Storage lipid accumulation by zygotic and somatic embryos in culture. Annals of Botany. 1989;64(4):409-420. DOI:10.1093/oxfordjournals.aob.a087761

[55] Chugh A, Khurana P. Gene expression during somatic embryogenesis-recent advances. Current Science. 2002;83:715-730.

[56] Laux T , Würschum T , Breuninger H. Genetic Regulation of Embryonic Pattern Formation. The Plant Cell. 2004; (suppl 1) S190-S202; DOI: 10.1105/tpc.016014

[57] Jürgens G, Mayer U, Busch M, Lukowitz W, Laux T. Pattern formation in the Arabidopsis embryo: a genetic perspective. Philosophical Transactions of the Royal Society B Biological Science 1995;350(1331):19-25. DOI: 10.1098/rstb.1995.0132.

[58] Ikeda M, Umehara M, Kamada H. Embryogenesis-related genes; its expression and roles during somatic and zygotic embryogenesis in carrot and *Arabidopsis*. Plant Biotechnology. 2006;23:153-161.

[59] Devic M. The importance of being essential: EMBRYO-DEFECTIVE genes in *Arabidopsis*, Comptes Rendus Biologies. 2008;331(10):726-736, DOI: 10.1016/j.crvi.2008.07.014.

[60] Meinke DW. Genome-wide identification of *EMBRYO-DEFECTIVE* (*EMB*) genes required for growth and development in *Arabidopsis*. New

Phytologist. 2020;226:306-325. DOI: 10.1111/nph.16071

[61] McCready K, Spencer V, Kim M. The Importance of TOR Kinase in Plant Development. Frontiers in Plant Science. 2020;11:16. DOI: 10.3389/fpls.2020.00016.

[62] Deprost D, Truong HN, Robaglia C, Meyer C. An *Arabidopsis* homolog of RAPTOR/KOG1 is essential for early embryo development. Biochemical Biophysical Research Communications. 2005;326(4):844-850. DOI: 10.1016/j.bbrc.2004.11.117

[63] Anderson GH, Veit B, Hanson MR. The *Arabidopsis* AtRaptor genes are essential for post-embryonic plant growth. BMC Biology. 2005;3:12. DOI: 10.1186/1741-7007-3-12

Chapter 5

Breeding Mustard (*Brassica juncea*) for Salt Tolerance: Problems and Prospects

Jogendra Singh, Parbodh Chander Sharma and Vijayata Singh

Abstract

Salt stress is currently one of the most critical factors, reducing agricultural production. Indian mustard (*Brassica juncea*) is a major oilseed crop in these areas. However, salt affects as much as 50–90% worldwide yield reduction. Salt tolerance is a very complex factor controlled by a number of independent and/or interdependent mechanisms and genetic modification that lead to many changes in physiology and biochemistry at the cellular level. The classical methods of plant breeding for salt tolerance involves the widespread use of inter and intraspecific variations in the available germplasm which is essential for any crop development program. This large germplasm is then tested under various salt levels in microplots, which is a quick, reliable, reproducible and inexpensive method of salt tolerance. Genotypes that have shown better indications of stress tolerance without significant yield reduction are considered to be tolerant and are also used as potential donor in the breeding programs. In this way, ICAR-Central Soil Salinity Research Institute (ICAR-CSSRI), Karnal developed and produced five varieties of Indian mustard that tolerate high salt namely, CS 52, CS 54, CS 56, CS 58 and CS 60 in the country, and many other high-quality pipeline lines exploration and development. These salt-tolerant species work better under conditions of salt stress due to various manipulations (physiology, genes and molecular level) to fight salt stress has led to detrimental effects. Recent molecular tools to add classical breeding systems to improve saline-tolerant mustard varieties in a short span of time, including the Marker Assisted Selection (MAS) and backcrossing, that have helped using simple sequence repeats (SSR) and single nucleotide polymorphisms (SNP) markers to identify quantitative trait loci (QTLs) that control the polygenic traits like tolerance of salt and seed yield.

Keywords: mustard, *Brassica juncea*, salinity, salt tolerance, antiporter, antioxidant genes, QTLs

1. Introduction

Globally the total area of saline soil is 397 million ha and 434 million ha of sodic soil. Of the 230 million irrigated fields, 45 million ha (19.5%) are salt-affected and almost 1500 million ha of arable agriculture, 32 million (2.1%) salt-affected [1]. Of the world's salt-affected land, an estimated 6.73 million ha are in India. In addition, arid and semi-arid areas are associated with salty groundwater, which must be used for irrigation, due to the unavailability or diversion of quality water outside for agricultural purposes. The use of this salty groundwater makes the soil unsuitable for growing crops. Salt stress is currently one of the most critical factors, reducing agricultural production (**Figure 1**).

Reclamation of these salt affected areas is of paramount importance to bring more and more areas under cultivation. This is necessary to enhance the food availability for feeding the burgeoning population in the country. Generally, there are three approaches being followed for the reclamation of these salt affected soils. Of these, the engineering solution is beyond the reach of resource poor farmers due to its prohibitive cost and community based application. The chemical amendment approach is generally followed by the farmers, even for which the subsidies are required to be provided by the Governmental agencies. Further, the smaller land holdings with the resource poor farmers also act as a deterrent in the adoption of these technologies. Thirdly, the biological reclamation approach, by developing salinity and alkalinity tolerant crops, is cost effective and is also economically feasible.

Vast literature is available on the effects of salinity on crop plants. Higher amount of salt reduces the productivity of many agricultural crops [2]; salt stress has three way effects that reduce water potential and cause ion toxicity and imbalances [3]. Salt stress affects other major processes such as germination, germination speed, root/shoot dry weight and the Na^+/K^+ in the root and shoot [4]. Hence, salt tolerance is important during the life cycle of any plants. Excessive salinity reduces the productivity of many agricultural crops [2], Salt stress has three fold effects which reduces water potential and causes ion imbalance and toxicity [3]. These studies have revealed the complex and polygenic nature of plant salt tolerance. Potential of genetic approach towards solving the problems of soil salinity and alkalinity is now widely recognized and this approach may more relevant to areas that often facing hard constraints of availability of resources. The existence of sufficient heritable variability may help for Genetic adaptation of crops to salinity which permits the identification and selection of salt tolerant strains and traits confer salt tolerance. Modern varieties have a relatively narrow genetic base and are poorly adapted to adverse environments such as salinity. However, endemic genotypes from problem environments may provide the basic germplasm for breeding salt tolerant varieties with acceptable yield potentials. Notwithstanding, the, genetic variations for salt tolerance among agricultural crops are very less, because most of the cultivated genotypes have been selected from normal environment where salt tolerant traits must have been gradually discarded, however variability for salt

Figure 1.
View of natural salt affected soil.

tolerance are similar in many wild progenitors due to where natural selection in response to salty habitats.

Large amount of variability is present amongst different crops with respect to their behavior under salt stress and this has also been documented. Further, variability is also available within a particular crop for their performance under varying rootzone salinity. The availability of such kind of variability in a crop is an essential requirement for the improvement in its salt tolerance character besides retaining or incorporating the other desirable beneficial characters. Further, the pool of variability in any crop can also be enhanced by subjecting them to mutagenic agents, which can further be screened for the desired characters. Screening whole plants and the large amount of germplasm available for a particular crop for salinity tolerance in the field situations is really time consuming, labour intensive and herculean task. Keeping these factors in view, rapid screening methodologies have also been developed for screening large number of germplasm for salinity tolerance under solution culture in laboratory conditions. When plant breeders are faced with a task of breeding crop varieties which are to be used under specific problem conditions, the criteria of selection is essential to any advancement which may be possible. In case of salt resistance, it would seem that it is essential to work hand to hand with the plant physiologists and soil scientists, in conditions which would make reliable selection possible and to determine if parameters can be developed which can make selection possible and effective. Further, without a concerted and concentrated research effort, problem such as breeding for salt tolerance cannot be effectively pursued.

Brassicas are an important group of edible oil and vegetable plants of the Brassicaceae family. The group has six cultivated species, namely, *Brassica campestris*, *Brassica nigra* and *Brassica oleracea* are diploid; *Brassica juncea*, *Brassica napus* and *Brassica carinata* are oligo-tetraploids (**Figure 2**).

Brassicas is the third most important edible oil source in the world, after soybean and palm, grown in more than 50 countries around the world. China, Canada, India, Germany, France, UK, Australia, Poland and the USA are the major producers of various varieties of Brassica. Globally, India accounts for 21.7% and production area 10.7% [1]. In India, oil-seed Brassicas are cultivated at about 2.3 million ha salt affected fields out of 6.9 million total cultivated area, which fall under the arid region, affected by varying levels of saline soils [5]. *B. rapa*, *B. napus* and *B. juncea* are mainly grown for oil and seed meal. The most serious effects of salt stress in Brassica are a decrease in crop height, size and yield and product quality [6]. Salt stress has significantly affected the lipid components of mustard seeds. With the increase in salt, the total amount of neutral lipids decreased significantly, while phosphor-lipids and glycol-lipids increased. The fatty acid profiles of whole, neutral and polar lipid fractions are severely affected. The dry weight of the plant decreased significantly in high salt levels (ECe 8 to 12 dS/m) and the maximum weight was observed in ECe 4 dS/m [7]. Brassica varieties showed a lower percentage of oil content in seeds under saline soil conditions (ECe = 13 dS/m and SAR = 12.70). It may be due to excessive absorption of toxins ions that interfere with metabolic processes.

In addition, unhealthy nutritional imbalances due to stress-induced nutrient uptake; depletion in the germination, chlorophyll and mineral ions slow down seed growth and early crop maturity under high salt intake can be attributed to reduced oil content [8, 9]. Further the salinity also significantly reduced net photosynthesis, stomatal conductance, water use efficiency and transpiration under during the formation of siliqua results in the greater yield loss [10].

Higher salt (EC > 12 dS/m); decreased the oil, protein and crude fiber content by 5–7%, 15–20% and 29–34% respectively, while the content of erucic acid increased by 12–17% [5]. However, its growth and productivity are greatly reduced by salt.

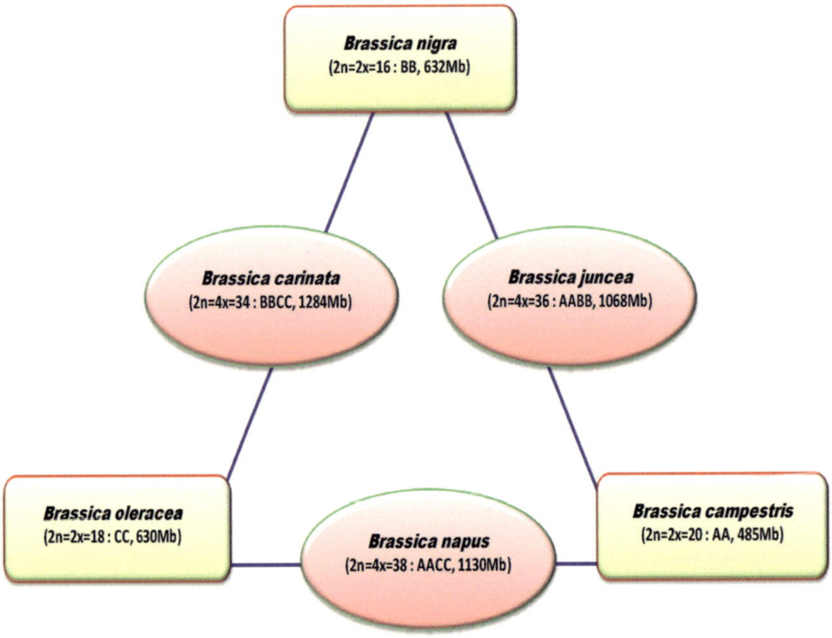

Figure 2.
The Triangle of U diagram shows the genetic relationship between the six species of the genus Brassica. Three of the Brassica species were derived from three ancestral genomes, denoted by the letters AA (campestris)*, BB* (nigra) *and CC* (oleracea)*. Alone each of these diploid genomes produced a common Brassica species.*

This situation can be mitigated in a way that includes water conservation and irrigation, crop management and crop production. There is a great deal of interest in the breeding stress-tolerant species, because significant genetic variations for salt tolerance exist between and within Brassica, which requires being exploited by selection and breeding. However, programs to develop salt tolerance species are hampered by traits complexity, inadequate genetic and physiological knowledge of tolerance-related factors, and a lack of an effective selection background. Improved mustard varieties with high salt tolerance and consumer accepted oil quality are required to achieve high yields and to increase the cultivated area under this stressful environment.

2. Development of salt tolerance in Brassicas

2.1 Germplasm characterization: right way to screen for salt tolerance

Salt tolerance is a complex characteristics that you can learn for the following reasons: (a) salt tolerance can only be tested under stressful conditions, which can affect many plant responses; (b) salt tolerance is a quantitative factor that requires effective and efficient methods of quantifying tolerance levels; (c) "salt" in "salt stress" is often misunderstood as it may contain different mineral salts, such as NaCl, $MgCl_2$, and $CaCl_2$; without excessive use of NaCl in salt, we cannot ignore the damage due to other ions; and (d) other abiotic stresses like drought, excess acidity and alkalinity, are often associated with salt exposed plants, making this difficult to study. Therefore, effective and efficient methods should be used, including plant culture under salt conditions, characterization and quantification of a salt tolerance level, in the first phase of the study.

Plants that grow under certain controlled conditions (e.g. hydroponics) are often used for salt tolerance studies because there is very little natural saline soil that can provide a representative and stable environment [11]. Large pots under the controlled conditions (Microplots/hydroponics) required for growth of Brassica plants and seedlings, while very less experiments for yield evaluation have been conducted in salt affected land. It is noteworthy that the salt tolerance of the Brassicas may be determined by a variety of genes, expressed by salt tolerance responses at various stages of development [12–15].

2.2 Control of salt stress environments

Diversification of locations, maximization of replications and monitoring of the environmental conditions during crop growth often provide a good control over the factors responsible for performance of a genotype or a set of genotypes. At ICAR-Central Soil Salinity Research Institute (ICAR-CSSRI), for large scale screening of varieties at germination and seedling stage, shallow-depth germination trays provided with a polythene sheet lining on the inner face are being used. They allow a simulation of germination response of the field nature, giving not only a quantitative indication of relative germination and survival rates but also the relative delays in germination, which is a characteristic of the different genotypes under salinity as well as sodicity stress. Apart from this, microplots of various sizes were constructed at the Institute filled with artificially prepared saline soil or original salty soil brought from salt affected fields, so that soil is uniform all through the profile. This way desired level of sodicity and salinity in these microplots can be maintained uniformly. Data obtained from microplots containing desired levels of saline or alkali soils, have been found to be well correlated with those collected from satisfactorily conducted field experiments. The field gradient of soil salinity is determined by soil tests at small intervals of space and a long strip running full length across the salinity/ sodicity gradient is allotted to each genotype. Further, irrigation with saline waters of predetermined composition is also practiced to establish desired soil salinity levels particularly when relative sensitivity of different growth stages are sought to be compared.

The genotypes with good germination rates has shown a reduction in fresh and dry weight in the vegetable phase under salt stress than in poorly developed ones. Therefore, salt tolerance trials throughout the life cycle or in areas where salt is most sensitive, will be required to compare salt tolerance in different lines [16]. Methods of artificial salt stress, such as slow compression and shock of salt, can lead to results different from those of field testing [17]. The enforcement of salt stress by the gradual exposure to NaCl instead of salt shock has been recommended in genetic and molecular studies because it reflects natural phenomena of salt stress. However, the ideal type of gradual salt impose is technically difficult [18, 19]. Researchers are looking for a simpler or more accurate approach to predicting salt tolerance so that they can better select tolerant plant species or tolerant genotypes. The ability to accumulate photosynthates, proline and glycine-betaine, as well as ion precipitation can be used as a means of biochemical or physiological selection for salt tolerance in canola [20, 21]. The accumulation pattern for various salt overly sensitive (SOS) transcripts after 24 hours of salt stress in various cultivars showed a strong positive association with salt tolerance among Brassica species [22]. Cell membrane stiffness associated with antioxidant enzyme activities (superoxide dismutase, catalase and peroxidase) can be particularly effective in identifying canola with high salt tolerance. To date, no uniform index has been used to test salt tolerance [23].

2.3 Development of salt tolerant cultivars: conventional methods

Breeding salt tolerance in crop plants is considered one of the ways to combat the global problem of increasing soil salinity in agricultural land. Stresses under adverse soil conditions are very complex and are often associated with climate hazards. The salt stress varies from place to place even during the season. Soil salinity is often associated with unhealthy nutrient inequalities (deficiencies/toxins) and other problems and plants adapted different types of strategies to overcome on it (**Figure 3**).

The interaction between soil salinity and other environmental factors influences the plant's response to that salt stress. Such problems are due to the slow evolution of plant species that thrive in adverse edaphic areas [24]. Therefore, it is necessary that the genetic material of plants should be tested in targeted areas with sufficient salt stress to find reliable sources of tolerance. Developing crop varieties with increased salt tolerance are considered to be the most promising, energy-saving and economical method than major engineering processes and soil rehabilitation techniques that have exceeded the limits of smallholder farmers [25].

2.3.1 The genetic basis of salt tolerance in Brassicaceae

Exploration of the heritable potential of a certain trait within the existing germplasm for a given crop would provide information on factors such as salt tolerance for plant breeders. The both additive and non-additive gene actions involved of in the inheritance of characteristics. High narrow-sense heritability estimates were observed for Ca^{2+}, K^+, Na^+, K^+/Na^+, Ca^{2+}/Na^+ and stress tolerance index, indicating the prime importance of additive effects in their genetic control [26]. Higher estimates of GCV, PCV, heritability and genetic advance (% of mean) under saline condition was observed for main shoot length, number of pods on main shoot and yield per plot, indicated that these characters might be controlled by additive genes [27, 28]. Salt tolerance was mainly controlled by dominant genes with an additive

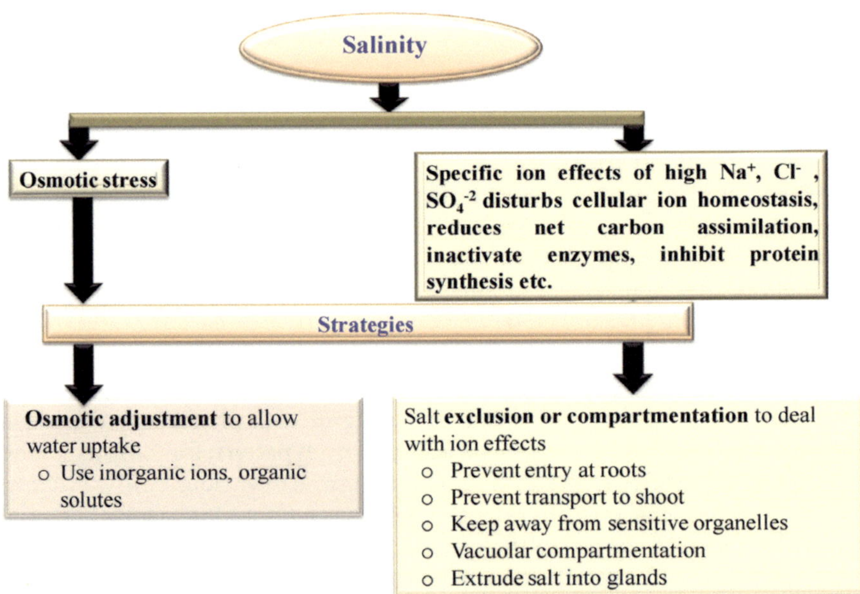

Figure 3.
Problems due to Salt stress and combating strategies in plants.

effect. The dominant effect played a major role and over-dominance might have existed in salt tolerance [29, 30]. The traits like main shoot length, number of pods on main shoot and yield per plot could be improved effectively by selection as these might be controlled by additive genes. Indian mustard, which was thought to be the moderately salt-tolerant species, also showed a decrease in shoot length and root length, electrolyte leakage, protein content, K^+/Na^+ ratio due to differential regulation of Na^+ in root and main stem by inhibition of entry from roots to shoot and retain higher photosynthetic characteristics than other species [10]. The fencing of selection processes should therefore be based on such indicators as a priority in the development of the most productive varieties of Indian mustard for saline condition.

In an effective breeding program, the discovery of a large variety of potential variants in a plant's genetic pool is a prerequisite; such genepools are needed to provide the required genetic diversity. Genetic diversity provides parental material from well-adapted landraces to enhance local adaptation. It helps to overcome the tendency to find a problem in the soil and provides a basis for fulfilling the needs of the novels. The conventional methods of improving plant salt tolerance generally employ selection for seed yield and there are few examples of producing salt tolerant varieties following these approaches at ICAR-CSSRI. These varieties are extremely popular with the farmers and their certified seeds are in great demand. The areas under their cultivation is fast expanding and increasing every year. The adoption of these varieties by the farmers has helped in great deal to enhance their economic status.

2.3.2 Bulk method

Using this methods of breeding researchers at the ICAR-Central Soil Salinity Research Institute (ICAR-CSSRI), Karnal has developed five cultivars of salt-tolerant Indian Mustard (*Brassica juncea*); CS 52, CS 54, CS 56, CS 58 and CS 60 (**Table 1**).

In this method, space planting of F_1 was done and harvested in bulk, while the planting of F_2 to F_6 generations done at commercial seed rate and spacing and harvested in bulk (**Figure 4**). The size of population in each generation was about 30,000 plants. These were space planted in the F_7 generation, and, only 5000 plants with desired characters confers to salt tolerance under salinity (ECe 12.0 dS/m) and sodicity (pH 9) conditions were selected. Seeds of these selected plants were separately harvested. Individual plant progenies were grown in multi-row plots. Weak and inferior progenies were rejected and only 300 individual homozygous plant progenies with desirable traits were selected and harvested in bulk. A preliminary yield trial was conducted for two years for agronomic traits and resistance/tolerance to disease and mustard aphid infestation, along with the national check varieties. Replicated yield trials were conducted for three years under saline and alkaline conditions in salt-affected soils [30].

2.4 Development of salt tolerant cultivars: non-conventional methods

If genetic diversity is fully utilized by continuous selection, then diversity may be sought through alternatives such as chemical and radiation, protoplast fusion, or recombinant DNA techniques. Different laboratories are undertaking studies on elucidating salt tolerance mechanisms following molecular and biotechnological approaches. Efforts for the sequencing of *Brassica juncea* genome is underway at different locations although a draft sequence has been published but a clear understanding of the agriculturally important traits is lacking. In the meantime, we are

Parameter/Variety	CS 52	CS 54	CS 56	CS 58	CS 60
Year of development	1997	2005	2008	2017	2018
Plant height (cm)	170–175	160–170	198–202	180–185	182–187
Maturity duration (days)	130–135	121–125	132–135	130–135	125–132
Seed type	Medium	Bold	Medium	Bold	Bold
1000-seed weight (g)	4.5–5.0	5.0–5.5	4.5–5.0	5.0–5.5	5.0–5.2
Salinity tolerance (ECe dS/m)	6–9	6–9	6–9	6–11	6–12
Sodicity tolerance (*p*H)	8.5–9.3	8.5–9.3	8.5–9.3	8.5–9.4	8.5–9.5
Yield in non stress(t/ha)	1.8–2.0	2.0–2.4	2.2–2.6	2.6–2.8	2.5–2.9
Yield in salt stress(t/ha)	1.5–1.6	1.6–1.9	1.6–1.9	2.0–2.2	2.0–2.2
Oil Content	37–38%	38–39%	38–39%	39–40%	40–41%
Time of sowing	Upto 15th October	Upto 15th October	Upto 15th November	Upto 25th October	Upto 25th October
Recommended ecology			Salt affected Areas		

Table 1.
Salinity tolerant cultivars of Brassica species developed through conventional breeding.

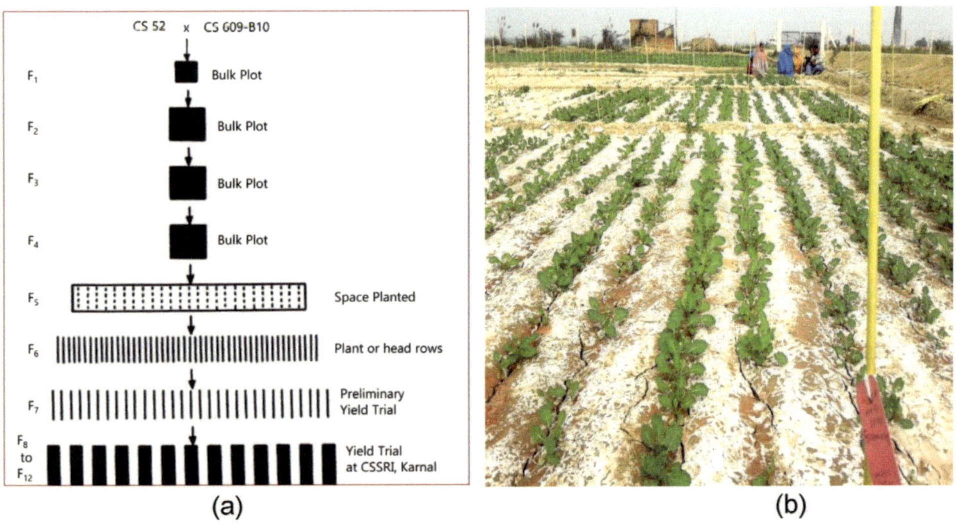

Figure 4.
Development of salt tolerant Indian mustard variety CS 60 (a) Bulk Method; (b) Genotype CS 60 under saline field (ECe 15 dS/m).

suggesting some studies that would help in further evaluating mustard germplasm for these traits through molecular techniques and will providing basis for development of salt tolerant brassica through non-conventional methods.

2.4.1 The molecular basis of salt tolerance in Brassicaceae

More recently, research into salt tolerance in plants has shifted from genetic mapping to molecular characterization of salt responsive genes. Increased understanding of biochemical pathways and mechanisms that involved in plant stress response has made it clear that many of these methods are common defense mechanisms that can be used by salt, drought and cold, although sometimes alternatives signaling pathways may be used. The molecular mechanism of salt tolerance expressed in model plants will facilitate the identification of target genes and the development of transgenic salt-tolerant plants in Brassica plants (**Figure 5**). Overexpression of antiporters (SOS1, SOS2, SOS3, ENH and NHX) as well as antioxidant genes (MPK1, DHAR3, APX1, APX4 and MDHAR6) in mustard play an important role in reducing the effects of salt and enhance salt tolerance [10].

The SOS pathway consists of the plasma membrane Na^+/H^+ antiporter SOS1, the protein kinase SOS2, and the Ca^{2+} binding protein SOS3. An increase of Na^+ concentration elevate the intracellular Ca^{2+}, and SOS3 binds Ca^{2+} and activates SOS2 to form a compound that phosphorylates membrane-derived plasma SOS1. Finally, over-expression of SOS1 leads to Na^+ efflux overhead [31]. In addition, AtHKT1 is involved in the recirculation of Na^+ from shoots to roots, possibly by promoting Na^+ movement into phloems in shoots and translocation into roots. The role of AtNHX1 in salt tolerance through increased Na^+ compartmentation in the vacuoles [32–35]. SOS1 and SOS3 are constitutively expressed in Brassica plants, while the pattern of SOS2 expression amongst *Brassica* species was found to be very unique. The expression of SOS2 may be elevated by salinity stress in the roots of all the *Brassica* species except for *B. juncea*, which maintains high SOS2 transcripts even under non-stress conditions, indicating a very unique feature of *B. juncea* [22]. Strong correlation between transcript abundance for SOS pathway related genes and salinity stress tolerance was observed in *Brassica* crops [36]. Currently, transgenic plants have been used to test the effect of overexpression of certain plant genes, which are known to be controlled by salt stress. Efforts have been made to increase transgenic Brassica with genetic predisposition, using genes that have a proven role in ion homeostasis, accumulating osmolytes, etc., to make them more tolerant of salt stress.

Transgenic *B. rapa* spp. chinensis plants that express the gene for choline oxidase (codA) from *Arthrobacter globiformis* have shown significantly higher net photosynthesis under high salinity conditions than wild-type plants [37]. The deception of these genes can help reduce the effects of ionic toxicity and cellular homeostasis as well as the conditioning of photosynthetic traits that lead to a promising yield under salt stress. Therefore, with the genetic improvement of agro-morphological characteristics of salt tolerance in Indian mustard, researchers should pay close

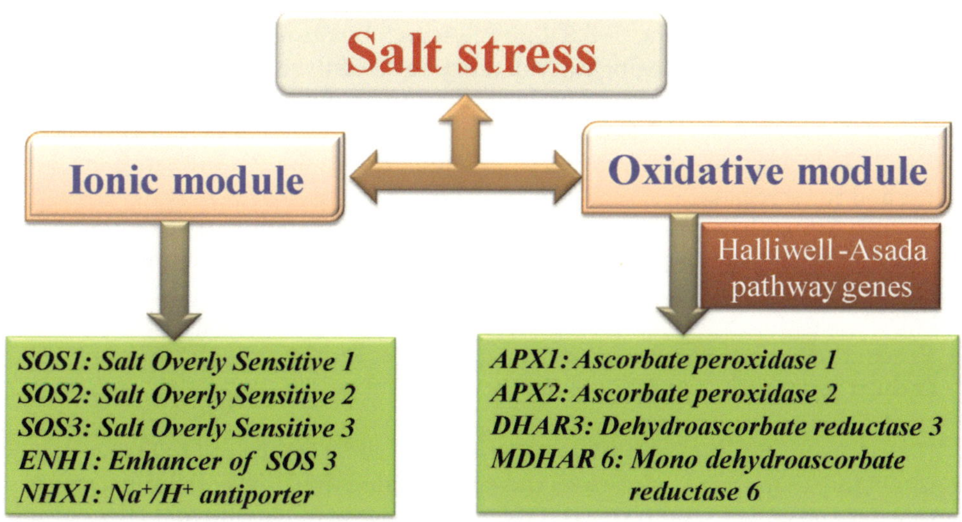

Figure 5.
The existence of a more efficient salt scavenging system composed of ionic module (SOS1, SOS2, SOS3, ENH and NHX) and oxidative module (MPK1, DHAR3, APX1, APX4 and MDHAR6) in the salt tolerant mustard.

attention to the photosynthetic attributes and pyramiding of antiporters and antioxidant genes for high economic productivity under salt stress. The overexpressing LEA4-1 plays an important role in the salt tolerance at vegetative stage in B. napus while BnLEA4-1 increase tolerance to salt stress in Arabidopsis [38]. Similarly Glutathione (GSH) and γ-ECS (Glutamylcysteine synthetase) gene from *B. juncea* (BrECS) plays an important role in cell function and metabolism as an antioxidant and provides plants with improved salt tolerance by maintaining the cellular nature of GSH redox to avoid attacks from salinity-derived reactive oxygen species [39].

2.4.2 Quantitative trait loci (QTLs) for salt tolerance

The QTL mapping is the best way to identify the underlying genes, though it is difficult and time-consuming. Creating an association map, which uses the highest number of historical recombination events/relics that occur throughout the evolutionary process of mapping population, enables genetic engineering in small genomic regions [40]. Exciting results have been obtained from independent studies on salt tolerance in the Brassicaceae, particularly in Arabidopsis. Most of the identified QTLs that control salt tolerance were different from each other, because the difference in mapping populations and the features under investigation. Normal QTL for germination percentage was detected at 20 cM in chromosome 1 associated with the RAS1 gene, a poor salt-tolerant controller during seed germination and early growth [41]. Another QTL found at 50 cM in chromosome 4 of the candidate AT4G19030 gene [42], whose level of expression reduced by ABA and NaCl [43]. These results suggest a complex genetic network regulating salt tolerance with differential genetic determinants in different accessions. Other QTLs of various traits are embedded: for example, salt responses and root-length QTLs on chromosomes 1 and 3, indicating that these two loci may contain gene-regulating salt tolerance expressed by root growth. However, genome-wide association studies with larger samples are considered to be more reliable and highly productive.

However, studies on QTLs or genes that regulate salt tolerance in Brassica plants are still very limited. To date, the practice of breeding salt tolerance in Brassica has been unsuccessful due to the unavailability of the polymorphic and cross transferability markers and highly salt sensitive lines. Concerns have resulted in a comprehensive breeding program for the development of high-yielding salt-tolerant mustard at the ICAR-Central Soil Salinity Research Institute (ICAR-CSSRI), Karnal and also leading to the changing salt tolerance paths of Brassica juncea by mutation results in the development of highly salt sensitive mutant CS 614-1-1-100-13 and CS 245-2-80-7 that are being used in recombinant inbred lines for mapping of QTLs. Researchers and farmers are trying to understand the salt-tolerance mechanisms and the screen for stable salt-tolerant genotypes to be used in the breeding programs. Efforts have also been made to develop salt-tolerant Brassica transgenic plants with a gene-specific role in ion homeostasis and osmolyte accumulation [44].

3. Predicted model for deciphering salinity tolerance mechanism in Indian mustard

Based on our findings on we have developed a model for the salt tolerance mechanism in Indian mustard (**Figure 6**) and conditioning the differential functions of antiporter and antioxidant transcripts in the mitigation of detrimental effect of salt stress [45]. Model suggested the three-way effect of salt stress on mustard plants; (i) Decreasing stomatal conductance results in the decreased intercellular CO_2 which caused diminishing activities of photosynthetic enzymatic machinery and

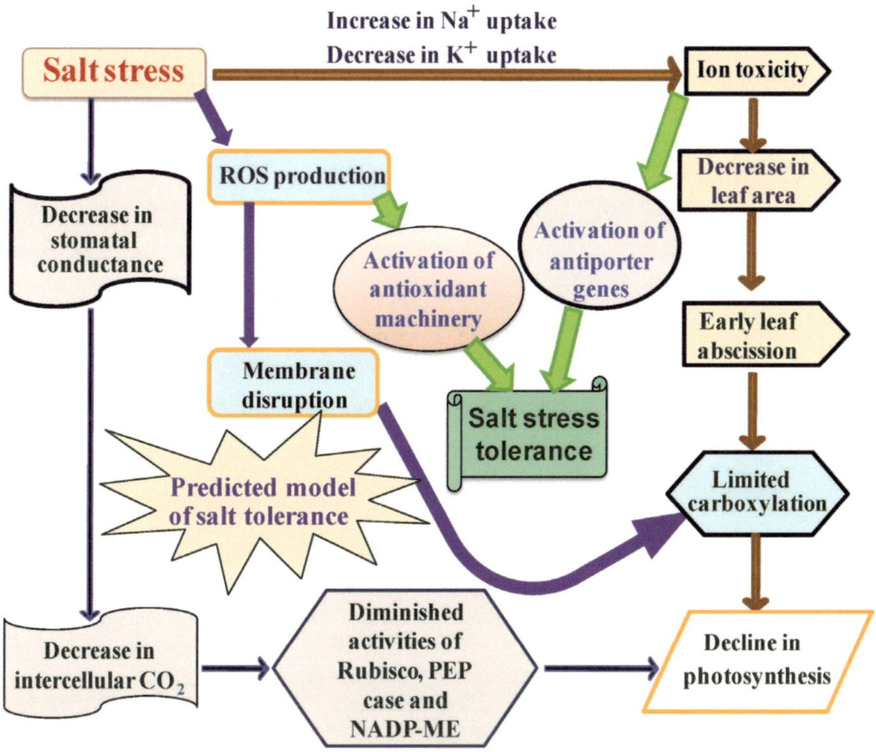

Figure 6.
A predicted model for the salt tolerance mechanism in Indian mustard.

decline in net photosynthesis rate. (ii) Production of reactive oxygen species (ROS) which disrupt the membrane system and limited the carboxylation process results in the least photosynthesis. (iii) Imbalance in the cellular ionic concentrations due to increased uptake of Na^+ and decreased K uptake which caused ion toxicity. This ion toxicity leads to decrease in leaf area and early leaf fall down and limited carboxylation results in declined photosynthesis rate. The salt tolerant mustard genotypes counteract on these toxic paths by activation of antioxidant gene network for ROS scavenging and antiporter gene complex that enhanced sequestration of Na^+ in roots and reduced toxic Na^+ transport to shoots, hence, makes mustard plant tolerant to salt stress.

4. The conclusion

Modern agriculture certainly requires commercial crops that tolerate salt for the purpose of crop trade. Genetic adaptation of crops to salinity requires that sufficient heritable variability exists within species to permit selection of salt tolerant strains and that those plant characteristics that confer salt tolerance be identified. Modern varieties have a relatively narrow genetic base and are poorly adapted to adverse environments such as salinity. However, endemic genotypes from problem environments may provide the basic germplasm for breeding salt tolerant varieties with acceptable yield potentials. Notwithstanding, the, genetic variations for salt tolerance among agricultural crops are very less, because most of the cultivated genotypes have been selected from normal environment where salt tolerant traits must have been gradually discarded, however variability for salt tolerance are similar in many wild progenitors due to where natural selection in response to salty habitats. Recent in-depth studies have identified various pathways at physiological and cell levels in which wild plants respond to salt stress. Due to the close relationship and significant variability between and within the Brassica species show great potential for breeding salt tolerance in Brassica plants. However, it is clear that to connect the salt tolerance factor and the QTL site to the chromosome, a proper breeding system assisted by markers is a prerequisite.

Conflict of interest

All the authors declare that they have no conflict of interest.

Author details

Jogendra Singh*, Parbodh Chander Sharma and Vijayata Singh
ICAR-Central Soil Salinity Research Institute, Karnal, Haryana, India

*Address all correspondence to: jogendra.singh@icar.gov.in

References

[1] FAO. Status of the World's Soil Resources (SWSR)–Main Report. Food and Agriculture Organization of the United Nations, Rome. 2015.

[2] Metternicht GI, Zinck JA. Remote sensing of soil salinity: Potentials and constraints. Remote Sensing of Environment. 2003;**85**:1-20

[3] de la Peña R, Hughes J. Improving vegetable productivity in a variable and changing climate. SAT eJournal. 2007;**4**(1):1-22

[4] Parida AK, Das AB. Salt tolerance and salinity effects on plant: a review. Ecotoxicology and Environmental Safety. 2005;**60**:324-349

[5] Singh J, Sharma PC, Sharma SK, Rai M. Assessing effect of salinity on oil quality parameters of Indian mustard (*Brassica juncea* L. Czern & Coss) using Fourier Transform Near-Infrared Reflectance (FT-NIR) spectros-copy. Grasas y Aceites. 2014; 65: e009.

[6] Zamani Z, Nezami MT, Habibi D, Khorshidi MB. Effect of quantitative and qualitative performance of four canola cultivars (*Brassica napus* L.) to salinity conditions. Advances in environmental biology. 2010;**4**:422-427

[7] Parti RS, Deep V, Gupta SK. Effect of salinity on lipid components of mustard seeds (*Brassica juncea* L.). Plant Foods for Human Nutrition. 2003;**58**:1-10

[8] Ali A, Mahmood IA, Salim M, Arshadullah M, Naseem AR. Growth and yield of different brassica genotypes under saline sodic conditions. Pakistan Journal of Agricultural Research. 2013;**26**:9-15

[9] Mahmood IA, Ali A, Shahzad A, Salim M, Jamil M, Akhtar J. Yield and quality of *Brassica* cultivars as affected by soil salinity. Pakistan Journal of Scientific and Industrial Research. 2007;**50**:133-137

[10] Singh J, Singh V, Vineeth TV, Kumar P, Neeraj, Sharma PC. Differential response of Indian Mustard (*Brassica juncea* L., Czern & Coss) under salinity: photosynthetic traits and gene expression. Physiology and Molecular Biology of Plants. 2019; 25(1):71-83.

[11] Flowers TJ. Improving crop salt tolerance. Journal of Experimental Botany. 2004;**55**:307-319

[12] Almodares A, Hadi MR, Dosti B. Effects of salt stress on germination percentage and seedling growth in sweet sorghum cultivars. Journal of Biological Sciences. 2007;**7**:1492-1495

[13] Ashraf M, Sharif R. Does salt tolerance varies in a potential oilseed *Brassica carinata* at different growth stages? Journal of Agronomy and Crop Science. 1998;**181**:103-115

[14] El Madidi SE, Baroudi BE, Aameur FB. Effects of salinity on germination and early growth of barley cultivars. International Journal of Agriculture and Biology. 2004;**6**:767-770

[15] Islam MM, Karim MA. Evaluation of rice genotypes at germination and early seedling stage for their tolerance to salinity. Agriculturists. 2010;**8**:57-65

[16] Quesada V, García-Martínez S, Piqueras P, Ponce MR, Micol JL. Genetic architecture of NaCl tolerance in *Arabidopsis*. Plant Physiology. 2002;**130**:951-963

[17] Shavrukov Y. Salt stress or salt shock: which genes are we studying? Journal of Experimental Botany. 2013;**64**:119-127

[18] Katori T, Ikeda A, Iuchi S, Kobayashi M, Shinozaki K, Maehashi K, et al. Dissecting the genetic control of natural variation in salt tolerance of *Arabidopsis thaliana* accessions. Journal of Experimental Botany. 2010;**61**:1125-1138

[19] Singh V, Singh AP, Bhadoria J, Giri J, Singh J, Vineeth TV, et al. Differential expression of salt-responsive genes to salinity stress in salt-tolerant and salt-sensitive rice (*Oryza sativa* L.) at seedling stage. Protoplasma. 2018;**255**(6):1667-1681

[20] Munir S, Hussain E, Bhatti KH, Nawaz K, Hussain K, Rashid R, Hussain I. Assessment of inter-cultivar variations for salinity tolerance in winter radish (*Raphanus sativus* L.) using photosynthetic attributes as effective selection criteria. World Applied Sciences Journal. 2013; 21:384-388.

[21] Ulfat M, Athar HR, Ashraf M, Akram NA, Jamil A. Appraisal of physiological and biochemical selection criteria for evaluation of salt tolerance in canola (*Brassica napus* L.). Pakistan Journal of Botany. 2007;**39**:1593-1608

[22] Kumar G, Purty RS, Sharma MP, Singla-Pareek SL, Pareek A. Physiological responses among Brassica species under salinity stress show strong correlation with transcript abundance for SOS pathway-related genes. Journal of Plant Physiology. 2009;**166**:507-520

[23] Ashraf M, Ali Q. Relative membrane permeability and activities of some antioxidant enzymes as the key determinants of salt tolerance in canola (*Brassica napus* L.). Environmental and Experimental Botany. 2008;**63**:266-273

[24] Ashraf M, Nazir N, McNeilly T. Comparative salt tolerance of amphidiploid and diploid *Brassica* species. Plant Science. 2001;**160**:683-689

[25] Rezai AM, Saeidi G. Genetic analysis of salt tolerance in early growth stages of rapeseed (*Brassica napus L.*) genotypes. Indian Journal of Genetics and Plant Breeding. 2005;**65**:269-273

[26] Sinha TS, Singh D, Sharma PC, Sharma HB. Genetic variability, correlation and path coefficient studies and their implications of selection of high yielding genotypes in Indian mustard under normal and sodic soil conditions. Indian Journal of Coastal Agricultural Research. 2002;**20**:31-36

[27] Kumar S, Mishra MN. Genetic variation and association analysis in Indian mustard (*Brassica juncea* (L.) Czem and Coss) and *Brassica campestris* L. variety toria. Crop Research. 2006;**31**(3):391-393

[28] Qiu Y, Li X, Zhi H, Shen D, Lu P. Genetic analysis of salinity tolerance in *Brassica campestris* ssp. *chinensis* (L.) Makino var. *communis* Tsen et Lee. Journal of Zhejiang University Science B. 2009;**10**(11):847-851

[29] Long WH, Pu HM, Zhang JF, Qi CK, Zhang XK. Screening of *Brassica napus* for salinity tolerance at germination stage. Chinese Journal of Oil Crop Sciences. 2013;**35**:271-275

[30] Singh J, Sharma PC, Singh V, Sharma DK, Sharma SK, Singh YP, et al. CS 58: new high yielding, salt and alkaline tolerant cultivar of Indian mustard. Crop Breeding and Applied Biotechnology. 2019;**19**(4):461-465

[31] Martinez-Atienza J, Jiang X, Garciadeblas B, Mendoza I, Zhu JK, Pardo JM, et al. Conservation of the salt overly sensitive pathway in rice. Plant Physiology. 2007;**143**:1001-1012

[32] Berthomieu P, Conéjéro G, Nublat A, Brackenbury WJ, Lambert C, Savio C, et al. Functional analysis of AtHKT1 in *Arabidopsis* shows that Na^+ recirculation by the phloem is crucial

for salt tolerance. EMBO Journal. 2003;**22**:2004-2014

[33] Apse MP, Aharon GS, Snedden WA, Blumwald E. Salt tolerance conferred by overexpression of a vacuolar Na^+/H^+ antiporter in *Arabidopsis*. Science. 1999;**285**:1256-1258

[34] Zhang HX, Blumwald E. Transgenic salt-tolerant tomato plants accumulate salt in foliage but not in fruit. Nature Biotechnology. 2001;**19**:765-768

[35] Zhang HX, Hodson JN, Williams JP, Blumwald E. Engineering salt-tolerant *Brassica* plants: characterization of yield and seed oil quality in transgenic plants with increased vacuolar sodium accumulation. Proceedings of the National Academy of Sciences, USA. 2001;**98**:12832-12836

[36] Chakraborty K, Sairam RK, Bhattacharya RC. Differential expression of salt overly sensitive pathway genes determines salinity stress tolerance in *Brassica* genotypes. Plant Physiology and Biochemistry. 2012;**51**:90-101

[37] Wang QB, Xu W, Xue QZ, Su WA. Transgenic *Brassica chinensis* plants expressing a bacterial codA gene exhibit enhanced tolerance to extreme temperature and high salinity. Journal of Zhejiang University Science B. 2010;**11**(11):851-861

[38] Dalal M, Tayal D, Chinnusamy V, Bansal KC. Abiotic stress and ABA-inducible Group 4 LEA from *Brassica napus* play a key role in salt and drought tolerance. Journal of Biotechnology. 2009;**139**:137-145

[39] Bae MJ, Kim YS, Kim IS, Choe YH, Lee EJ, Kim YH, et al. Transgenic rice overexpressing the Brassica *juncea* gamma-glutamylcysteine synthetase gene enhances tolerance to abiotic stress and improves grain yield under paddy field conditions. Molecular Breeding. 2013;**31**:931-945

[40] Nordborg M, Tavaré S. Linkage disequilibrium: what history has to tell us. Trends in Genetics. 2002;**18**:83-90

[41] Ren Z, Zheng Z, Chinnusamy V, Zhu J, Cui X, Iida K, et al. RAS1, a quantitative trait locus for salt tolerance and ABA sensitivity in *Arabidopsis*. Proceedings of the National Academy of Sciences, USA. 2010;**107**:5669-5674

[42] Lee BH, Kapoor A, Zhu J, Zhu JK. STABILIZED1, a stress-upregulated nuclear protein, is required for pre-mRNA splicing, mRNA turnover, and stress tolerance in *Arabidopsis*. Plant Cell. 2006;**18**:1736-1749

[43] deRose-Wilson L, Gaut BS. Mapping salinity tolerance during *Arabidopsis thaliana* germination and seedling growth. PLoS ONE. 2011; 6:e22832.

[44] Zhang JZ, Creelman RA, Zhu JK. From laboratory to field. Using information from Arabidopsis to engineer salt, cold, and drought tolerance in crops. Plant Physiology. 2004;**135**:615-621

[45] Singh J, Sharma PC, Singh V, Kumar P. Predicted model to reveal the mechanism of salt tolerance in *Brassica juncea*. Journal of Soil Salinity and Water Quality. 2019;**11**(1):18-30

Chapter 6

Salinity Tolerance in Canola: Insights from Proteomic Studies

Ali Bandehagh, Zahra Dehghanian, Robert Henry and Mohammad Anwar Hossain

Abstract

Salinity considerably lowers crop yield worldwide. Production of salt stress-tolerant species will be essential to maintain the food supply in the coming decades. Brassicas, including various members of the family Brassicaceae, are very necessary sources of human food. Importantly, the key crop species that are members of the Brassicaceae family are genetically diverse and therefore their response reaction and adaptation to salinity varies greatly. Canola (*Brassica napus* L.) is commonly grown for edible oils and other uses such as biodiesel fuel production. Although most types of canola are identified as salt-resistant, plant yield and development are reduced significantly by rising salinity levels. In saline situations, the plant's genome supports a range of physiological changes in some plant characteristics. Since the function of genes cannot indicate the exact condition of cells, proteomic approaches are emerged as methods to investigate the plant's responses to stresses in the molecular levels. Exploring the proteome complements research at the genome and transcriptome level and helps elucidate the mechanism of salt tolerance in plants. Proteins are reliable indicators of salinity responses, as they are directly involved in forming the new phenotype providing adaptation to salinity. In this chapter, we review the response of the rapeseed proteome to salinity stress.

Keywords: canola, plant proteomics, molecular markers, ROS, salinity

1. Introduction

Plants developing under field conditions are exposed to many ecological factors, which define their macro and micro environment. Every deviance in these environmental variables from the optimal levels may be detrimental to the plant and cause stress. Stress is provided by abiotic factors such as elevated drought, salinity, temperature, or biotic factors such as bacteria, insects, nematodes, fungi, and viruses. Plants may also have to cope with multiple stresses. Among the abiotic stresses, salinity has emerged as one of the most important extreme agents which limit agricultural crop yield as well as making large areas unsuitable for cultivation. About 7 percent of the world regions, 20 percent of all the world's agricultural lands, and nearly half of the irrigated lands are impacted by soil salinity [1]. In addition, the areas impacted by salt increase about 10 percent annually and if the problem is not solved, about 50 percent of the arable lands will be salinized by the year 2050 [2]. Soil salinity has substantial negative effects on the growth and productivity of plants. Reduction of plant growth and productivity can result from a disproportionate supply of photosynthetic assimilates or hormones to growing tissues [3].

The lack of cultivable areas due to salinization is a direct challenge to providing the burgeoning population with adequate supplies of food. Therefore, there is a need to grow species that, are not only able to endure high levels of salt but can also retain optimum yields levels in the presence of salt. Nonetheless, due to the multigenic and quantitative nature of salt tolerance, endeavors to improve crop production under salinity have been generally ineffective. This has motivated researchers to follow a combination of approaches utilizing both traditional and novel strategies to improve salt tolerance.

Members of the Brassicaceae are major contributors to the daily needs of humans and are used as vegetables, oils, condiments and more. Brassica members occupy the third position in vegetable oil production between various oilseed species [3, 4]. *Brassica napus* L. or as we know it, Canola, is commonly grown for edible oils and biological-based fuel production. Notwithstanding many types of canola being regarded as salt-resistant, plant development and yield decrease in response to rising salinity stress. Morphological characteristics such as shoot and root development during stress; physiological factors such as water of leaf content, chlorophyll amount, photosynthetic amount, and membrane integrity; biochemical agents such as osmolytes accumulation, the activity of antioxidant enzymes and molecular responses such as modifications of expression of salt-resistant genes are key traits for detecting and characterizing differences in resistance within and between members of the Brassicaceae family [5].

Acclimatization to stress is mediated by deep modifications in the expression of genes that lead to variations in the plant transcriptome, metabolome, and proteome. Many investigations have shown that changes in the transcript-level expression of genes do not always lead to changes in protein levels [6, 7]. Hence it is also necessary to examine changes in the proteome because proteins are direct agents of plant response to stress. Furthermore, proteins include enzymes that catalyze modifications in the amounts of metabolites, and also regulatory proteins, for example they may control the plant response to stress at the transcription and protein translation levels.

Proteins contribute to stress-acclimation mechanisms which are directed to modifications in the cell cytoplasm, cytoskeleton, plasma membrane, and combination of the intracellular compartment that include alteration in their effects, for example, cell cytoplasm affinity to water [6, 7]. Modifications in protein accumulation under stress are firmly linked to the phenotypic reaction of the plant resisting the stress. Studies of plant reactions to stress at the protein level may contribute significantly to our understanding of the processes of plant tolerance or resistance.

2. Salinity

Saline soils are among the main environmental constraints that limit plant growth owing to the high salt concentration and the process of incremental increase in salt content is known as salinization [1]. Soil salinization is a global issue and often affects the littoral zone by extending soil salinity [8]. Almost 7% of the whole global land area, 20% of the cultivated area, and about half of the irrigated area is affected by soil salinity [1]. Furthermore, salt-affected area is increasing by 10% yearly per year and more than 50% of arable land will be salinized by the year 2050 if the problem is not addressed [9].

2.1 The salinity types and causes

2.1.1 Primary or natural salinity

Salinity stress occurs as a result of salt deposition by natural processes in the soils or groundwater over a long time period. Two normal mechanisms are involved. The first one is the weathering of parental substrates containing dissolved salts. The processes of weathering decompose rock and release dissolved salts of different forms, primarily sodium chloride (NaCl), with sulfates, magnesium, calcium, and carbonates, in lower concentrations. NaCl is the most abundant form of dissolved salt. The second mechanism is the deposition of maritime salt sediments which is transported by rain and wind. "Cyclic salts" are maritime salts which are transported by wind and sedimented by precipitation.

2.1.2 Human-induced or secondary salinity

Secondary salinization is the consequence of human actions that alter the soil's hydrological equilibrium between water used (irrigation or precipitation) and water used by crop [10]. The most important factors are (i) field clearance and the replacement of perennial vegetation annuals and (ii) irrigation, with salt-rich water and without adequate drainage.

2.2 Salinity stress effects on canola development and productivity

Soil salinity, similar to drought stress, is as a major abiotic stress which causes reduced crop production globally [11]. Growth and development of many plant species, including Brassica species, are adversely affected when subjected to salinity stress, due to the restriction of essential physiological, biochemical and metabolic processes with the consequence of toxicity of ions, osmotic stress, and decreased supply of water and minerals [12]. Salinity decreases nutritional ions like Fe, Zn, and Mn levels in plant organs including leaves, stems, and roots and pods at the flowering phases in Brassica spp. [13]. Plant height is reduced due to salinity stress and caused by decreased osmotic, leaf water potential, and enhancing electrolyte leakage [14].

Salinity stress adversely affects germination of canola seeds [15], reduces the length of radicles and plumes and seedling fresh weight, decreases biomass [16–18], impairs seed filling stage and the number of pods on plants, decreases the number of seeds in each pod and pod length [19], reduces the number of leaves, flowers, branches and siliques, leads to shorter siliques, less seed per silique and 1000-weight of seed [15], decreases leaf size, leaf nutrient attraction levels, hypocotyl and root development in seedlings with a rise in IAA oxidase and activity of peroxidase enzyme [17, 20], decreases chlorophyll a, chlorophyll b and total chlorophyll [21] and also reduces total fatty acids by 25% [18].

The decline in chlorophyll concentration as salt increases causes lower dry weight and decreases average leaf weight but in salt-resistant canola cultivars, this drop in the leaf weight and height of plant does not happen [22]. Proline is accumulated in the roots of salt-resistant canola cultivars and the shoots of salt-sensitive cultivars [23].

In canola genotypes, it has been reported that the potassium (K^+) ion concentration diminish as salinity increased, while calcium (Ca^{2+}) and sodium (Na^{2+}) ion concentrations increased, decreasing the photosynthetic rate [24]. An incrementing Na^+ and the ratio of Na^+ to K^+ in shoot and root is found under salinity stress [25].

The aggregation of ions of Na^+ and chlorine (Cl^-) raises the osmotic potential and reduces the supply of water and the plant roots' nutrient absorption [26]. Toxic metabolic Na^+ ions compete with K^+ in many major physiological processes in cells [27]. Electrolyte leakage, which rises in salinity responses, is thus attributed to a rise in metabolites and ion concentrations and is correlated with an increase in the input of Cl^- and Na^+ and the omission of K^+ [28]. This leads to a considerable decrease in the shoot and root dry weight, leaf number, and shoot height under stress [29].

When subjected to biotic and abiotic stress, plants generate an abundance of antioxidant enzymes. Islam et al. [30], found that by affecting water and nutrient equilibrium, high salt concentrations in the root region, impair canola and mustard growth and yield. Lower stomatal conduction, nutrient absorption, more ion toxicity, and a misbalance in nutrient accessibility are the main reasons for decline in seed yield in Brassica spp. under salinity stress [3]. ROS-inhibitors, SOD or Superoxide dismutase, Catalase (CAT), Monodehydroascorbate reductase (MDHAR), Glutathione reductase (GR), and decreased glutathione (GSH) accumulation were greater in leaves of five salinity stress ed. canola varieties than in non-stressed plants [31]. While salt increased levels of MDA, hydrogen peroxide (H_2O_2), and phenolics were also seen in canola which was sensitive to salt. High cellular H_2O_2 concentrations were accumulated with lower MDA levels in salt-resistant canola plants [32]. There were increased amounts of chlorophyll, carotenoids, flavonoids, proline, and dissolved protein in the *Brassica napus* L. lab plantlets developed in the presence of SA and NaCl [33]. Oilseed brassicas have broad salinity stress resistance, amid these adverse effects, helping them to reconcile to a wide range of biological and environmental conditions [3].

3. Proteomic approaches

Proteomic technology exploits advances in protein isolation and protein recognition relying on mass spectrometry. This technology supports the study of tolerance mechanisms and plant reactions to abiotic stresses including salinity [34]. In 1996, Marc Wilkins [35] introduced the phrase 'proteome' for the first time, a term which is at present associated with 133,606 publications in the proteomics field as presently listed in the NCBI database, of which 15,642 publications are associated with proteome/proteomics with stress studies and only 543 publications report proteome/proteomics studies associated with plant salinity stress [36]. Proteomics alludes to the large-scale and expansive study of all the proteins (the protein equivalent of the genome) to discover cellular processes [37]. Systematic proteome studies provide information on protein abundances, protein changes, and modifications, as well as interacting protein partners and protein networks [38].

In genotypes that are susceptible to salinity stress, the plant proteome is differentially expressed. Proteomics has a wide range of applications in protein profiling analysis under stress conditions. It has a direct role in the discovery of genes and proteins involved in plant salinity stress response and tolerance processes [39]. The introduction of genes encoding proteins, for the synthesis of the osmolytes, receptors, ion channels, and salt-responsive signaling factors or enzymes into salt-sensitive plants, can confer salinity-tolerant phenotypes [40]. High-throughput proteomics is the first step in characterization of salt-responsive proteins that can be used to produce salt-tolerant plants.

Another application is in comparative study of differential expression of proteomes among control (non-stressed) plants and stressed plants. Less often, the comparison of proteomes isolated from two variant genotypes or plant species with different extreme levels of salinity stress is studied. The proteomes are distinguished focusing on both protein quality and quantity by differential-expression in proteomic studies, which aim at both protein identification and relative quantitation [41].

Many experiments relevant to the comparative analysis of proteomes among plants exposed to salinity stress and control treatments have been conducted in economic plants such as, rice [42], *Brassica napus* [43], wheat [44], barley [45], tomato [46], soybean [47], and the model plant *Arabidopsis thaliana* and medicinal plants such as, *Andrographis paniculata* [48], *Bruguiera gymnorhiza* [49].

4. The goal of proteomics

Proteomic technologies are being more commonly used in many areas of bioscience, in addition to stress-responsive proteins detection in stress-tolerant plants, such as in the discovery of cell surface markers/biomarkers, and the production of drugs [50]. The target of proteomics is to provide complete information by revealing the regulation, amount, activities and interplay of proteins existing in complex biological systems, whole organism, specific tissues or cellular compartments in certain conditions and at a particular time [51]. Proteomics has become useful in the field of plant genomics in recent years, and may be used to identify proteins extracted from tissues/cells in response to growth and specific environmental conditions and to determine the levels of expression of the proteins found [52].

Researchers can ultimately evaluate and recognize thousands of proteins in each experiment with the application of these procedures. The relative expressions of these different proteins can be accurately determined and evaluated in different situations, and the expression of individual proteins may be appraised in intricate mixes [53]. Under stress conditions, functional identification of every protein and its metabolic processes, the protein profiles or mapping of cells, tissues, organ or organisms is valuable in the recognition of genes and gene product that are resistant to various stresses [54].

5. Reactive oxygen species (ROS) role in salinity

Salinity impacts plants by causing multiple problems including, ion toxicity, osmotic stress, nutritional deficit and genotoxicity resulting in the accumulation of ROS via oxidative stress. Salinity can result in stomatal closure, that decreases the supply of CO_2 in the leaves and prevents carbon fixation, exposing chloroplasts to extreme energy, that increases ROS development including H_2O_2, superoxide (O_2^-), singlet oxygen and hydroxyl radicals (OH-) [55, 56]. Since salinity is complicated and inflicts a water deficiency due to osmotic effects on an extensive range of metabolic processes [56]. This water shortage results in the formation of ROS [57]. ROS is extremely reactive and, by oxidation of lipids, proteins, and nucleic acid can cause cell harm [58]. In several reports, ROS manufacture has been shown to increase in salinity situations. ROS mediated membrane harm has been reported as an important factor in the toxicity to cells of salinity stress in many crops, including corn, tomatoes, citrus, peas, and pepper [55, 59, 60–62]. The increased function of GSH-Px and GR reduces the amount of H_2O_2 and MDA and ameliorates the impact on the plant (*Brassica napus* L.) by preventing oxidative harm stimulated by ROS under soil salinity stress [63].

Long term treatments with salinity, with EC 5.4 and 10.6 dS m − 1, for 60 days, have been shown to induce a substantial rise in H_2O_2 and lipid peroxidation in seedlings of wheat, which has more salt-sensitive cultivars than salt-resistant cultivars [64]. Lipid peroxidation improves the permeability of H_2O_2 and increases in these two compounds have been observed in *Brassica napus* [65] and *Triticum aestivum* [66] with increasing salinity. It has been shown that one of the key reasons for declines in crop productivity is the generation of ROS during environmental stresses such as salinity [67]. ROS control is thus a critical mechanism for preventing undesirable cellular cytotoxicity and oxidative harm [68].

Rehman et al. [69] recorded a 2.5- and a 3-fold increase in H_2O_2 generation, along with a 2- and 3-fold increase in the content of thiobarbituric acid reactive substances (TBARS) under 100 and 200 mM NaCl stress respectively, in contrast to controls demonstrating salt-induced oxidative stress. Oxidative stress differs between plant tissues. For example, root tissues have reported as suffered most from oxidative stress caused by salinity, followed by mature and young leaves.

6. Antioxidant defense system for salt tolerance

Antioxidants extirpate ROS directly or indirect and/or regulate ROS production [70]. The antioxidant defensive mechanism comprises non-enzymatic antioxidants of low molecular weight and certain enzymes acting on antioxidants [71]. In order to prevent hyper production of ROS, non-enzymatic antioxidants including ascorbate (AsA), GSH, tocopherol, phenolic combinations (PhOH), flavonoids, alkaloids, and nonprotein amino acids work in a coordinated manner with enzymes including superoxide dismutase (SOD), CAT, peroxidase (POX), (PPO) polyphenol oxidase, ascorbate peroxidase (APX), MDHAR, dehydroascorbate reductase (DHAR), GR, glutathione peroxidase (GPX), glutathione S-transferase (GST), TRX, and PRX [70].

The catalytic action of enzymes and non-enzymatic antioxidants and the locations of activity in the cellular organs is well known. The SOD is directly relevant to plants under salinity stress and begins the first phase of defense, transforming oxygen into hydrogen peroxide [72, 73]. The H_2O_2 generated may be further transformed into H_2O with the activity of enzymes; APX, CAT, GPX, or in the AsA-GSH cycle. Aggregation of SOD has been reported as a defensive approach in canola in response to salinity [43]. The cycle of AsA-GSH or the Asada-Halliwell cycle in plant cells is the main antioxidant protective pathway to detoxification of hydrogen peroxide, consisting of non-enzymatic antioxidants GSH and AsA and four major enzymes; DHAR, MDHAR, GR, and APX [74]. In the defense system of antioxidants, the main activity is executed by the AsA-GSH cycle to decrease H_2O_2 and redox homeostasis [74]. In the leaves of five canola cultivars under salinity tension, ROS-scavenging enzymes (SOD; CAT; GR; MDHAR), and in addition reduced glutathione concentration were more in unstressed leaves [75]. An increase in APX function in response to salinity stress is reported in *Brassica napus* [63].

A crucial function in the antioxidant defense mechanism is the reduction of H_2O_2 and redox homeostasis via the AsA-GSH cycle [76]. Furthermore, for detoxifying H_2O_2 and xenobiotics, GPX and GST are essential enzymes [77]. GSH and AsA are plentiful soluble antioxidants among the non-enzymatic antioxidants in higher plants, which act in a critical role as electron doners and directly remove ROS via the GSH-AsA cycle [76]. Furthermore, beta carotene reacts with ROO radicals, OH, and O_2 leading to decreased concentrations of cellular ROS [78].

It has been observed that Selenium (Se) increases the activity of these antioxidant enzymes to deal with established stresses [79]. Selenium plays an important role in various enzymatic and non-enzymatic processes for example phytochelatins and antioxidants of GSH, which helps defeat the salt-induced mass production of ROS. It has been proven that low amounts of selenite (Na_2SeO_4) protect plants from ROS-stimulated oxidative detriment, but a higher Se amount, acts as a pro-oxidant and promotes ROS production and oxidative stress [80]. Several researchers have identified the need for Se to improve ROS scavenging activity, decreasing MDA amounts, and membrane harm [81]. Reduced production of H_2O_2 has also been reported under increased Se [82]. Reduced H_2O_2 amounts were reported - under salinity stress in Se-treated canola (*Brassica napus* L.) plants [83]. Plants subjected to Se display less MDA under salinity conditions, demonstrating that Se is important in bringing down the lipid peroxidation by modifying the antioxidant enzymes and preserving the membrane structures of, *Brassica napus* L. [63]. Moreover, it has observed that the generation of lipid peroxidation (MDA) is decreased by increasing the amount of Se under salt conditions [84]. Reducing the fluidity of the membrane to increase membrane leakage and prevent harm to membrane proteins, ion channels, and enzymes are general effect of MDA on plant cell [85].

7. Heat shock proteins due to salinity stress

Crop breeding aims to enhance tolerances to salinity and high temperature. Organisms which survive in difficult conditions should have unique mechanisms to respond to stressful environments. One of these process would involve the induction of molecular chaperones, heat shock proteins (HSP), comprising some guarded protein families including HSP90, HSP70 (DnaK), HSP100 (ClpB), HSP60 (GroEL), and small heat shock proteins [86]. Studies of HSPs [87] have indicated that that sti1 (protein) was up-regulated in tolerance to salinity tension; this protein includes two heat shock chaperon binding motifs STI1), three tetratrico peptide repeats (TPR), and two Sti1 domains [88].

HSP90 is thought to interact with TPR-containing proteins via protein–protein interplay to modulate various cellular processes. HSP 70 has been verified in macro algae and some water plant species as a stress biomarker generated by NaCl, emphasizing its function in supporting species against stresses [89]. In order to image the entire modifications in the cells protein synthesis in tolerance to osmotic stress, dual channel imaging and warping of 2-DE protein gels have also been used. Analysis reveals that in many busy cellular surroundings, various chaperones adopt different paths to prevent protein aggregation. In canola, families of HSP have been identified in the leaf [43]. The differential expression of Hsp 70 has been reported in the root [43, 90].

It has been shown that transgenic plants expressing Hsp70 modulate programmed cell death (PCD) under hyper salinity, where Hsp70 functions as an anti-apoptotic protein [91]. In addition, ClpB/Hsp100 B2, B3 and ClpD2 are expected to act as molecular chaperones, and their expressions are significantly boosted under salt conditions [92]. Increased expression of ClpD1 and sHSP has also been shown to contribute to improved adaptation to salinity stress [93].

8. Salinity stress proteins as molecular markers

Plants under salinity stress change their gene expressions significantly to adapt to unfavorable conditions, including variations in the composition of the plant transcriptome, proteome, and metabolome. A few experiments have reported showing that protein aggregation varies considerably under stress condition [40, 92].

It is suggested that in canola root, Ras-related small GTP-binding proteins interfere with signaling of salinity stress. Proteomic investigations of canola cultivars under salinity stress have identified this protein [94]. The activation of Ras-related small GTP-binding proteins is responsible for coupling ligand-bound G proteins with GPCRs, which in turn sets the signaling pathway for Ca^{2+} mediated inositol triphosphate (IP3) in tolerance to salinity stress in canola leaves [95]. In transgenic *B. napus* plants, the probable mechanisms of activity of *ThIPK2* is an aggregation of sodium ions in the root, with differential expression of proline amount and stress-response genes [96].

The identity of a small GTP-binding protein Ras-relevant that is up-regulated in saline conditions in canola, indicates a high possibility of G-protein-couple recipients (GPCRs) being involved in intervention in detecting salinity signals [94]. This has clearly indicated that GPCRs, in combination with G-proteins, activate the small GTP-binding protein [97]. This step is accompanied by IP3 signaling pathway activation, generation of Ca^{2+}, activation of the Ca^{2+} process, and ultimately alters gene expression [98]. In connection with the function of the IP3 mechanism in the response of canola to salt conditions, it has been reported that certain components of the IP3 pathway are induced by high salinity. A *Brassica napus* transcriptomic study showed that phosphatidylinositol-specific phospholipase C2 (BnPLC2), phosphatidylinositol 3-kinase (BnVPS34), and phosphatidylinositol synthase (BnPtdIns S1) have substantially differential expression under salinity stress [99]. Annexin recognition in canola root supports this process in the detection and signaling of salinity stress in the case of the Ca^{2+} mechanism [94]. In responding to abiotic stress, the annexin mediator activities have been defined as goals of the Ca^{2+} signaling pathway [100]. The recognition of calcium-dependent protein kinase (CPK) differential expression under abiotic stress, like salinity stress, provides further verification of the active function of the above pathways in canola [101]. CPKs detect Ca2C and function as a kinase. Altogether, the sense of salinity stress is by GPCRs and Ras-related small GTP-binding proteins, transmitting the message via the IP3 signaling pathway.

9. Regulation of gene expression

In response to salinity in canola cultivar, three layers of adjustment of gene expression have been reported. The first phase of expression of gene adjustment is at the level of transcription, which is regulated by agents of transcription. In the regulatory regions of the genome, the transcription factor is an important factor that is interrelating with some other proteins, particularly RNA polymerases and also trans or cis components. Lee et al. [102], reported that fifty-six genes that encode transcription factors in canola are changed under abiotic stresses. In resistance to salinity the message is conveyed by messaging and sensing molecules, and these transcription factors are triggered. The expression of various genes is then regulated by diverse gene regulation networks composed of transcription factors and other proteins.

Another process related to gene adjustment, which has been observed in canola, is epigenetic activities. Epigenetic adjustment of stress-tolerate genes under varying situations has been found to perform a crucial task in the plant [103]. When salinity stress is added to a pre-treated plant with osmotic stress, histone changes aggregate Na^{+} ions in a concentration that is not toxic for the plants [104]. The methylation of DNA and changes of histones in reaction to salinity have been reported in canola. If the plant is subjected to salinity stress, de novo methylation and demethylation processes happen at CpCpGpG sites [105]. The main components of epigenetic adjustment are DNA methylation, histone changes, and chromatin reconstitution [106]. Most genes which have epigenetic alteration have minor identified in the plant (canola). A significant gene undergoing methylation of DNA under salinity stress in the plant is the ethylene-responsive element-binding factor (EBF) [107].

Studies on canola are very limited in this respect. It has been shown in Arabidopsis and tobacco that histone proteins are quickly up-regulated under salinity stress and are phosphorylated, due to reduced Na^+ aggregation [104]. The potential functions of DNA methylation/demethylation and chromatin (histone) modifications in adjusting the expression of salt-responsive genes are indicated by these findings.

Lu et al. [108] found that both hypermethylation and hypomethylation in the rapeseed genome were prompted by saline stress, and hypermethylation was observed more often than hypomethylation. There is a significant role of DNA methylation in plants reaction to abiotic stress [103]. Research has shown that salinity can influence the level of DNA methylation, and the shift in the status of cytosine methylation is associated with the variations in expression in rapeseed of two stress-related genes [109].

miRNA activities have been investigated in salinity treated canola at the post-transcriptional level. miRNAs are small, single-stranded, RNA molecules of 20–24 nucleotides which regulate the aggregation and expression of mRNAs. Multiple processes like organ growth [110], phase transfer [111], stress reactions [112], and many other regulative mechanisms for plants are indirectly modulated [113]. It has been reported that more than 340 miRNAs participate in the post transcription process of adjustment of the salt-responsive genes in canola [114]. The NAC transcription factor is one of the transcription factors found to be targeted by miRNAs [115]. NACs TFs are particular TFs with a strongly conserved N-terminal NAC domain and a variable CT activation domain. The function of NAC TFs in the tolerance of abiotic stress is well known [116]. Sixty NAC TFs have been reported in *B. napus* [116]. Zhong et al. [117] detected two *B. napus* NAC TFs (BnNAC2 and BnNAC5) and found that these factors act in negative adjustment of salinity and osmotic stress tolerance.

10. Dynamic variations of the genes and proteins of canola

Many transcriptomic and proteomic research conducted under salinity stress on canola suggest that differentially expressed proteins and genes can be predominantly grouped into seven functional groups in both leaves and roots [118]. The categories, with the proteins or genes characterized in each functional group, are (a) energy metabolism and carbohydrates, (b) defenses and stresses, (c) photosynthesis (in the leaves) and metabolism, (d) structure of cells, (e) transport and membrane, and (f) division of cell, fate, and differentiation [119].

The amount of protein relevant to carbohydrates and energy metabolism is higher in canola roots under stress compared to other functional protein groups. Proteins linked to the metabolism of amino acids and the composition of cells are significant in abundance. The bulk of the proteins are from the TCA cycle, the electron transport chain, and glycolysis of carbohydrates and energy metabolism [120]. In canola leaves, the highest abundance functional proteins are those that belong to photosynthesis, the degradation and synthesis of proteins, metabolism of amino acid, and damage repair and defense response [121, 122].

In the photosynthesis-related salinity-tolerant canola cultivars, differential redundancy of chlorophyll a/b binding protein, chloroplast RuBisCO activase, ribulose bisphosphate oxygenase/carboxylase (RuBisCO) both subunits, have been found [43, 90, 118]. It appears that under salinity stress, canola changes the cytoskeleton essential ingredients (actions and tubulins). The dynamic remolding of the cytoskeleton, like the K^+ channel, is related to certain of the major transmembrane transports [40, 123]. Another interesting point that is related to the functional group of differentially altered proteins is the unknown ones that form about 1% to 20% of total diversely altered proteins in every research outcome, particularly in researches on the root [43, 122]. The discovery of the function of these proteins could offer further insight into the path ways of salt response [124].

CDPKs, which are sensor responders with the capability to self-modifying verification by the enzymatic function is the third portion of the Ca^{2+} sensing machinery in the plant [125]. This causes CPKs to be special in their calcium-sensing dual function and then respond against the stress situation signals by downstream phosphorylation activities. In the stress response of CPKs, tremendous overlapping and cross-talk have reported [126]. There are several CPKs necessary for the reaction to a particular stress stimulus against stresses like drought, heat, salt, and cold. In *B. napus*, 25 CPKs have recognized and many have studied their expression levels under different abiotic stresses [126]. A study [127] of BnCPK2 interacting partners has been reported using a mating based split ubiquitin system (mbSUS) and BiFC. To control ROS and cell death, they suggested the role of BnCPK2 and probable interactions with NADPH oxidase-like respiratory burst oxidase homolog D (RbohD). Similar results have been obtained in which most CPKs are shown to dampen ABA signals and ROS homeostasis in the plant cell [128].

11. Suspected genes/proteins responsible for canola resistance

Multiple experiments has been conducted to recognize the major gene(s)/protein(s) responsible for salt tolerance. Understanding the main components of the salt response pathways complexities is an important step towards the production salt-tolerant canola. A study [121] identified 6 genes (hub) in resistant cultivars, including malate dehydrogenase, heat shock protein 70, triose phosphate isomerase, fructose-bisphosphate aldolase, UDP-glucose dehydrogenase and methionine synthase in the produced protein–protein interaction pathway of canola salt-induced proteins. Hub genes are highly interactive components of the response network that are known to be the network's core components [129]. Some of the suspected genes or proteins for canola resistance can be derived from research on the usage of materials that boost salinity tolerance of canola. Garg and Manchanda [85] observed, in response to plant growth promoting rhizobacteria inoculation, canola roots control glyceraldehyde-3-phosphate dehydrogenase and downregulates S-adenosylmethionine synthase, aldehyde dehydrogenase, and malate dehydrogenase under 150 and 300 mM of NaCl.

This research showed, inoculated plants exhibit substantially increased root dry weight, root length, more potassium, and less sodium and chlorine amounts compared to non-inoculated plants. The research found that the greater development of the inoculated roots was the reason for the differentially abundant bacteria-responsive proteins. As suggested in other research, inoculation with bacteria grants canola greater resistance via an enhanced redundancy of proteins which are related to glycolysis, and amino acid metabolism, TCA, and succinate dehydrogenase [90].

A few reports have revealed that overexpression of certain genes contributes to changed salinity tolerance in canola. The Dehydration-Responsive Element Binding transcription Factors (DREBs) overexpression is a major example of this. Plants transformed for great expression of DREBs show a discrete increase in their salinity tolerant expression of gene like HSF3, COR14, RD20, HSP70, and PEROX, which indicate greater resistance. The plants, which are transgene, can live in the saline condition, where wild species plants are more susceptible [130].

Considering the exogenous use of 5-Aminolevulinic acid (5-ALA) resulted in salt tolerance, in treated plants, Sun et al. [131] transformed canola with the

5-ALA-encoding gene, YHem1, and studied the growth of the transgenic canola, ability to synthesize further 5-ALA, and wild-type canola under salt conditions. They observed that under both short-term and long-term salt conditions, transgenic canola demonstration more product, more chlorophyll amount, a greater amount of antioxidant enzyme, high proline content, high sugar content, and more free amino acids contrasting to wild-type canola. They also have shown improved resistance of transgenic canola may be linked to the up-regulation of the Rubisco small subunit and a substantial amount of Fe metal. In comparison to these experiments, in which improved resistance has been documented, it has proposed that expression of *Brassica napus* TTG2 induces sensitivity to salinity stress through the down-regulation of the Tryptophan Biosynthesis 5 (TRP5) and YUCCA2 (YUC2) indole-3-acetic acid (IAA) encoding genes, thus decreasing the endogenous IAA amounts. In future research in transgenic plants, the recently evolving CRISPR/Cas9 system is expected to provide more knowledge on molecular components that respond to salinity stress [132].

12. Conclusions and prospects of proteomics

Canola as a major field crop across the world is influenced by salinity stress. Despite advances in understanding molecular interactions between plant and salt, production of salinity-resistant cultivars remains challenging. Proteomics findings help greatly to identify the physiological processes based on plant tolerance to stress, and could further be used to identify the level of stress tolerance in genotypes. To date, we have substantial information gaps concerning the regulation of abiotic stress plant response, as this adjustment is at different levels of transcription, post transcription, post-translation, and epigenetic levels [133].

A variety of stress acclimation techniques have been explained in the above studies through a combination of proteomics and physiological approaches. Nevertheless, many of the findings demonstrated the previously characterized salt-induced proteins rather than offering new mechanistic insights into salinity tolerance. More knowledge on alterations in cell metabolism and also stress-responsive proteins, which are participating in the proteome of plant, are provided whereas there is a lack of knowledge about regulating proteins in stress, expression of gene regulation and signaling proteins (mainly transcription factors), membrane proteins and transferors, owing to their limited amount in the cell or difficulty in characterization.

In the near future, the advent of new improved proteomic techniques and the study of unique developmental stages/cells/tissues or subcellular organelles, in particular using LCM (Laser Capture Mediated Micro Dissection)-mediated single-cell isolation, will allow the study of cell-specific expression, protein enrichment and low-abundant protein detection to be successfully achieved [134].

Author details

Ali Bandehagh[1*], Zahra Dehghanian[2], Robert Henry[3]
and Mohammad Anwar Hossain[4*]

1 Department of Plant Breeding and Biotechnology, Faculty of Agriculture, University of Tabriz, Tabriz, Iran

2 Department of Biotechnology, Faculty of Agriculture, Azarbaijan Shahid Madani University, Tabriz, Iran

3 Queensland Alliance for Agriculture and Food Innovation, University of Queensland, Brisbane, Australia

4 Department of Genetics and Plant Breeding, Bangladesh Agricultural University, Mymensingh, Bangladesh

*Address all correspondence to: bandehhagh@tabrizu.ac.ir
and anwargpb@bau.edu.bd

References

[1] Shrivastava P. and Kumar R. Soil salinity: a serious environmental issue and plant growth promoting bacteria as one of the tools for its alleviation. Saudi journal of biological sciences. 2015;22(2): 123-131.

[2] Jamil A. Riaz S. Ashraf M. and Foolad M.R. Gene expression profiling of plants under salt stress. Critical Reviews in Plant Sciences. 2011; 30(5): 435-458.

[3] Ashraf M. and McNeilly T. Salinity tolerance in Brassica oilseeds. Critical reviews in plant Sciences. 2004;23(2): 157-174.

[4] Purty RS, Kumar G, Singla-Pareek SL, Pareek A. Towards salinity tolerance in Brassica: an overview. Physiology and Molecular Biology of Plants. 2008;14(1-2): 39-49.

[5] Mondal S. and Chakraborty K. Brassicaceae Plants Response and Tolerance to Salinity, The Plant Family Brassicaceae. Springer. 2020; pp. 203-228.

[6] Gygi SP, Rochon Y, Franza BR, Aebersold R. Correlation between protein and mRNA abundance in yeast. Molecular and cellular biology. 1999;19(3): 1720-1730.

[7] Bogeat-Triboulot MB, Brosché M, Renaut J, Jouve L, Le Thiec D, Fayyaz P, Vinocur B. Gradual soil water depletion results in reversible changes of gene expression, protein profiles, ecophysiology, and growth performance in Populus euphratica, a poplar growing in arid regions. Plant physiology. 2007;143(2): 876-892.

[8] Gupta B. and Huang B. Mechanism of salinity tolerance in plants: physiological, biochemical, and molecular characterization. International journal of genomics. 2014.

[9] Zhang S, Fan C, Wang Y, Xia Y, Xiao W, Cui X. Salt-tolerant and plant-growth-promoting bacteria isolated from high-yield paddy soil. Canadian journal of microbiology. 2018; 64(12): 968-978.

[10] Manchanda G. and Garg N. Salinity and its effects on the functional biology of legumes. Acta Physiologiae Plantarum. 2008;30(5): 595-618.

[11] Yamaguchi T. and Blumwald E. Developing salt-tolerant crop plants: challenges and opportunities. Trends in plant science. 2005; 10(12): 615-620.

[12] Hasegawa PM, Bressan RA, Zhu JK, Bohnert HJ. Plant cellular and molecular responses to high salinity. Annual review of plant biology. 2000;51(1): 463-499.

[13] Chakraborty K, Sairam R, Bhaduri D. Effects of different levels of soil salinity on yield attributes, accumulation of nitrogen, and micronutrients in *Brassica* spp. Journal of plant nutrition. 2016;39(7): 1026-1037.

[14] Dolatabadi N. and Toorchi M. Rapeseed (*Brassica napus* L.) genotypes response to NaCl salinity. Journal of Biodiversity and Environmental Sciences. 2017;10: 265-270.

[15] Mahmoodzadeh H. Comparative study of tolerant and sensitive cultivars of *Brassica napus* in response to salt conditions. Asian Journal of Plant Sciences. 2008.

[16] Bybordi A. and Tabatabaei J. Effect of Salinity Stress on Germination and Seedling Properties in Canola Cultivars (*Brassica napus* L.). Notulae Botanicae Horti Agrobotanici Cluj-Napoca. 2009;37(2).

[17] Bybordi A, Tabatabaei SJ, Ahmadev A. Effect of salinity on

the growth and peroxidase and IAA oxidase activities in canola. Journal of Food, Agriculture and Environment. 2010a;8(1): 109-112.

[18] Bybordi A, Tabatabaei SJ, Ahmadev A. Effects of salinity on fatty acid composition of canola (*Brassica napus* L.). Journal of Food, Agriculture and Environment. 2010b;8(1): 113-115.

[19] Mohammadi S, Shekari F, Fotovat R, Darudi A. Effect of laser priming on canola yield and its components under salt stress. International Agrophysics. 2012;26(1).

[20] Shainberg I. and Shalhevet J. Soil salinity under irrigation: Processes and management, 51. Springer Science & Business Media. 2012.

[21] Bahrani A. Effect of salinity on growth, ions distribution, accumulation and chlorophyll concentrations in two canola (*Brassica napus* L.) cultivars. American-Eurasian Journal of Agricultural & Environmental Sciences. 2016;13(5): 683-689.

[22] Kamrani M, Hosseinniya H, Chegeni A. Effect of salinity on the growth characteristics of canola (*Brassica napus* L.). Technical Journal of Engineering and Applied Sciences. 2013;3(18): 2327-2333.

[23] Saadia M, Jamil A, Akram NA, Ashraf M. A study of proline metabolism in canola (*Brassica napus* L.) seedlings under salt stress. Molecules. 2012;17(5): 5803-5815.

[24] Rameeh V. Ions uptake, yield and yield attributes of rapeseed exposed to salinity stress. Journal of soil science and plant nutrition. 2012;12(4): 851-861.

[25] Parida AK. and Das AB. Salt tolerance and salinity effects on plants: a review. Ecotoxicology and environmental safety. 2005; 60(3): 324-349.

[26] Acosta-Motos JR, Ortuño MF, Bernal-Vicente A, Diaz-Vivancos P, Sanchez-Blanco MJ. Hernandez JA. Plant responses to salt stress: adaptive mechanisms. Agronomy. 2017;7(1): 18.

[27] Tester M. and Davenport R. Na+ tolerance and Na+ transport in higher plants. Annals of botany. 2003;91(5): 503-527.

[28] Mohammadreza S, Amin B, Forogh A, Hossin M, Sorayya S. Response of *Brassica napus* L. grains to the interactive effect of salinity and salicylic acid. Journal of Stress Physiology & Biochemistry. 2012;8(2).

[29] Bandehhagh A, Toorchi M, Mohammadi A, Chaparzadeh N, Salekdeh GH Kazemnia H. Growth and osmotic adjustment of canola genotypes in response to salinity. Journal of Food Agriculture and Environment.2008;6(2): 201.

[30] Islam M, Bhuiyan M, Prasad B, Quddus M. Salinity effect on yield and component characters in rapeseed and mustard varieties. Journal of Biological Sciences. 2001;1(9): 840-842.

[31] Bybordi A, Tabatabaei SJ, Ahmadev A. The influence of salinity stress on antioxidant activity in canola cultivars (*Brassica napus* L.). Journal of Food, Agriculture & Environment. 2010c;8(1): 122-127.

[32] Rasheed R, Ashraf MA, Parveen S, Iqbal M, Hussain I. Effect of salt stress on different growth and biochemical attributes in two canola (*Brassica napus* L.) cultivars. Communications in soil science and plant analysis. 2014;45(5): 669-679.

[33] Razavizadeh R. Protein pattern of canola (*Brassica napus* L.) changes in response to salt and salicylic acid in vitro. Biological Letters. 2015;52(1-2): 19-36.

[34] Qureshi MI, Qadir S, Zolla L. Proteomics-based dissection of stress-responsive pathways in plants. Journal of plant physiology. 2007;164(10): 1239-1260.

[35] Wilkins MR, Pasquali C, Appel RD, Ou K, Golaz O, Sanchez JC, Yan JX. From proteins to proteomes: large scale protein identification by two-dimensional electrophoresis and arnino acid analysis. Bio/technology, 1996;14(1): 61-65.

[36] NCBI. https://pubmed.ncbi.nlm.nih.gov. Accessed 15 June, 2020.

[37] Pandey A. and Mann M. Proteomics to study genes and genomes. Nature. 2000;405(6788): 837-846.

[38] Park OMK. Proteomic studies in plants. BMB Reports. 2004;37(1): 133-138.

[39] Wang L, Liu X, Liang M, Tan F, Liang W, Chen Y, Lin Y. Proteomic analysis of salt-responsive proteins in the leaves of mangrove Kandelia candel during short-term stress. PLoS One. 2014;9(1): e83141.

[40] Tuteja N. Mechanisms of high salinity tolerance in plants. In 'Osmosensing and osmosignaling'. Elsevier Academic Press: San Diego, CA. 2007.

[41] Kosová K, Vítámvás P, Prášil IT, Renaut J. Plant proteome changes under abiotic stress—contribution of proteomics studies to understanding plant stress response. Journal of proteomics. 2011;74(8): 1301-1322.

[42] Nyong'a TM, Yang P, Li M. Proteomics Study in Rice Responses and Tolerance to Salt Stress, Advances in Rice Research for Abiotic Stress Tolerance. Elsevier. 2019. p. 781-789.

[43] Bandehagh A, Salekdeh GH, Toorchi M, Mohammadi A, Komatsu S. Comparative proteomic analysis of canola leaves under salinity stress. Proteomics. 2011;11(10): 1965-1975.

[44] Ma Q, Shi C, Su C, Liu Y. Complementary analyses of the transcriptome and iTRAQ proteome revealed mechanism of ethylene dependent salt response in bread wheat (*Triticum aestivum* L.). Food Chemistry. 2020;126866.

[45] da Silva Júnior WF, Bezerra de Menezes DL, De Oliveira LC, Koester LS, Oliveira de Almeida PD, Lima ES, De Azevedo EP. Inclusion Complexes of β and HPβ-Cyclodextrin with α, β Amyrin and In vitro Anti-Inflammatory Activity. Biomolecules. 2019;9(6): 241.

[46] Tang H, Zhang X, Gong B, Yan Y, Shi Q. Proteomics and metabolomics analysis of tomato fruit at different maturity stages and under salt treatment. Food Chemistry. 2020;311: 126009.

[47] Yin Y, Yang R, Han Y, Gu Z. Comparative proteomic and physiological analyses reveal the protective effect of exogenous calcium on the germinating soybean response to salt stress. Journal of proteomics. 2015;113: 110-126.

[48] Hossain MS. The Effect of Salinity Stress on the Morpho-physiology and Protein Profile of *Andrographis paniculata*, Kulliyyah of Science, International Islamic University Malaysia. 2016.

[49] Tada Y. and Kashimura T. Proteomic analysis of salt-responsive proteins in the mangrove plant, *Bruguiera gymnorhiza*. Plant and Cell Physiology. 2009;50(3): 439-446.

[50] Abdallah C, Dumas-Gaudot E, Renaut J, Sergeant K. Gel-based and gel-free quantitative proteomics approaches at a glance. International journal of plant genomics. 2012.

[51] Plomion C, Lalanne C, Claverol S, Meddour H, Kohler A, Bogeat-Triboulot MB, Barre A. Mapping the proteome of poplar and application to the discovery of drought-stress responsive proteins. Proteomics. 2006;6(24): 6509-6527.

[52] Aslam B, Basit M, Nisar MA, Khurshid M, Rasool MH. Proteomics: technologies and their applications. Journal of chromatographic science. 2017;55(2): 182-196.

[53] Andrecht S. and von Hagen J. General aspects of sample preparation for comprehensive proteome analysis. Proteomics sample preparation. Weinheim: Wiley-VCH Verlag GmbH & KGaA. 2008;5-20.

[54] Witzel K, Weidner A, Surabhi GK, Börner A, Mock HP. Salt stress-induced alterations in the root proteome of barley genotypes with contrasting response towards salinity. Journal of experimental botany. 2009;60(12): 3545-3557.

[55] Ahmad P, Jaleel C, Sharma S. Antioxidant defense system, lipid peroxidation, proline-metabolizing enzymes, and biochemical activities in two *Morus alba* genotypes subjected to NaCl stress. Russian Journal of Plant Physiology. 2010a;57(4): 509-517.

[56] Cheeseman JM. Mechanisms of salinity tolerance in plants. Plant physiology. 1988;87(3): 547-550.

[57] Elstner EF. Metabolism of activated oxygen species, Biochemistry of metabolism. Elsevier; 1987. p. 253-315.

[58] Ahmad P, Jaleel CA, Salem MA, Nabi G, Sharma S. Roles of enzymatic and nonenzymatic antioxidants in plants during abiotic stress. Critical reviews in biotechnology. 2010b;30(3): 161-175.

[59] Dionisio-Sese ML. and Tobita S. Antioxidant responses of rice seedlings to salinity stress. Plant science. 1998;135(1): 1-9.

[60] Parida AK. and Das A.B. Salt tolerance and salinity effects on plants: a review. Ecotoxicology and environmental safety. 2005;60(3): 324-349.

[61] Ahmad P. and Sharma S. Salt stress and phyto-biochemical responses of plants. Plant Soil Environ. 2008;54(3): 89-99.

[62] Ahmad P, Nabi G. and Ashraf M. Cadmium-induced oxidative damage in mustard [*Brassica juncea* (L.) Czern. & Coss.] plants can be alleviated by salicylic acid. South African Journal of Botany. 2011;77(1): 36-44.

[63] Hasanuzzaman M, Hossain MA. and Fujita M. Selenium-induced up-regulation of the antioxidant defense and methylglyoxal detoxification system reduces salinity-induced damage in rapeseed seedlings. Biological Trace Element Research. 2011; 143(3): 1704-1721.

[64] Sairam RK, Rao KV, Srivastava G. Differential response of wheat genotypes to long term salinity stress in relation to oxidative stress, antioxidant activity and osmolyte concentration. Plant science. 2002;163(5): 1037-1046.

[65] Hasanuzzaman M. and Fujita M. Exogenous silicon treatment alleviates salinity-induced damage in *Brassica napus* L. seedlings by up-regulating the antioxidant defense and methylglyoxal detoxification system. Plant Biology, American Society of Plant Biology (abstr.). 2011a

[66] Hasanuzzaman M. and Fujita M. Selenium pretreatment upregulates the antioxidant defense and methylglyoxal detoxification system and confers enhanced tolerance to drought stress in rapeseed seedlings. Biological Trace Element Research. 2011b;143(3): 1758-1776.

[67] Asada K. Production and action of active oxygen species in photosynthetic tissues. Causes stress and amelioration of defense systems in plants. 1994; 77-104.

[68] Halliwell B. and Gutteridge J. Free radicals in biology and medicine, 2nd edn. Clarendon. Oxford. 1989.

[69] Rehman S, Abbas G, Shahid M, Saqib M, Farooq ABU, Hussain M, Murtaza B. Effect of salinity on cadmium tolerance, ionic homeostasis and oxidative stress responses in conocarpus exposed to cadmium stress: Implications for phytoremediation. Ecotoxicology and Environmental Safety. 2019; 171: 146-153.

[70] Carocho M. and Ferreira IC. A review on antioxidants, prooxidants and related controversy: natural and synthetic compounds, screening and analysis methodologies and future perspectives. Food and chemical toxicology. 2013;51: 15-25.

[71] Hasanuzzaman M, Bhuyan M, Anee TI, Parvin K, Nahar K, Mahmud JA, Fujita M. Regulation of ascorbate-glutathione pathway in mitigating oxidative damage in plants under abiotic stress. Antioxidants. 2019;8(9): 384.

[72] Laxa M, Liebthal M, Telman W, Chibani K, Dietz KJ. The role of the plant antioxidant system in drought tolerance. Antioxidants. 2019;8(4): 94.

[73] Del Río Sánchez C. Guía de ética profesional en psicología clínica. Comercial Grupo ANAYA, SA. 2018.

[74] Fotopoulos V, Ziogas V, Tanou G, Molassiotis A. Involvement of AsA/DHA and GSH/GSSG ratios in gene and protein expression and in the activation of defence mechanisms under abiotic stress conditions, Ascorbate-glutathione pathway and stress tolerance in plants. Springer; 2010. p. 265-302.

[75] Bybordi A, Tabatabaei SJ. and Ahmadev A. The influence of salinity stress on antioxidant activity in canola cultivars (*Brassica napus* L.). Journal of Food, Agriculture & Environment. 2010; 8(1): 122-127.

[76] Hasanuzzaman M, Bhuyan M, Mahmud J, Nahar K, Mohsin S, Parvin K, Fujita M. Interaction of sulfur with phytohormones and signaling molecules in conferring abiotic stress tolerance to plants. Plant Signaling & Behavior. 2018;13(5): e1477905.

[77] Foyer CH. and Noctor G. Ascorbate and glutathione: the heart of the redox hub. Plant physiology. 2011;155(1): 2-18.

[78] Kapoor D, Singh S, Kumar V, Romero R, Prasad R, Singh J. Antioxidant enzymes regulation in plants in reference to reactive oxygen species (ROS) and reactive nitrogen species (RNS). Plant Gene. 2019;19: 100182.

[79] Zong Xj, Li Dp, Gu Lk, Li Dq, Liu Lx. and Hu Xl. Abscisic acid and hydrogen peroxide induce a novel maize group C MAP kinase gene, ZmMPK7, which is responsible for the removal of reactive oxygen species. Planta. 2009; 229(3): 485.

[80] Sobahan MA, Akter N, Ohno M, Okuma E, Hirai Y, Mori IC, Nakamura Y. Effects of exogenous proline and glycinebetaine on the salt tolerance of rice cultivars. Bioscience, biotechnology, and biochemistry. 2012; 76 (8):1568-1570.

[81] Shalaby T, Bayoumi Y, Alshaal T, Elhawat N, Sztrik A. and El-Ramady H. Selenium fortification induces growth, antioxidant activity, yield and nutritional quality of lettuce in salt-affected soil using foliar and soil applications. Plant and Soil. 2017; 421(1-2): 245-258.

[82] Hawrylak-Nowak B. Comparative effects of selenite and selenate on

growth and selenium accumulation in lettuce plants under hydroponic conditions. Plant Growth Regulation. 2013; 70(2): 149-157.

[83] Hashem HA, Hassanein RA, Bekheta MA. and El-Kady FA. Protective role of selenium in canola (*Brassica napus* L.) plant subjected to salt stress. Egypt J Exp Biol (Bot). 2013; 9(2): 199-211.

[84] Astaneh RK, Bolandnazar S, Nahandi FZ. and Oustan S. Effects of selenium on enzymatic changes and productivity of garlic under salinity stress. South African Journal of Botany. 2019; 121: 447-455.

[85] Garg N. and Manchanda G. ROS generation in plants: boon or bane? Plant Biosystems. 2009; 143(1): 81-96.

[86] Boston RS, Viitanen PV, Vierling E. Molecular chaperones and protein folding in plants. Plant molecular biology. 1996;32(1-2): 191-222.

[87] Chen L, Hamada S, Fujiwar, M, Zhu T, Thao NP, Wong HL, Krishna P. The Hop/Sti1-Hsp90 chaperone complex facilitates the maturation and transport of a PAMP receptor in rice innate immunity. Cell Host & Microbe. 2010; 7(3): 185-196.

[88] Takahashi A, Casais C, Ichimura K, Shirasu K. HSP90 interacts with RAR1 and SGT1 and is essential for RPS2-mediated disease resistance in Arabidopsis. Proceedings of the National Academy of Sciences. 2003; 100(20): 11777-11782.

[89] Ireland HE, Harding SJ, Bonwick GA, Jones M, Smith CJ, Williams JH. Evaluation of heat shock protein 70 as a biomarker of environmental stress in *Fucus serratus* and *Lemna minor*. Biomarkers. 2004;9(2): 139-155.

[90] Banaei-Asl F, Farajzadeh D, Bandehagh A, Komatsu S. Comprehensive proteomic analysis of canola leaf inoculated with a plant growth-promoting bacterium, *Pseudomonas fluorescens*, under salt stress. Biochimica et Biophysica Acta (BBA)-Proteins and Proteomics. 2016;1864(9): 1222-1236.

[91] Hoang TM, Moghaddam L, Williams B, Khanna H, Dale J. and Mundree SG. Development of salinity tolerance in rice by constitutive-overexpression of genes involved in the regulation of programmed cell death. Frontiers in plant science. 2015; 6: 175.

[92] Muthusamy SK, Dalal M, Chinnusamy V. and Bansal KC. Differential regulation of genes coding for organelle and cytosolic ClpATPases under biotic and abiotic stresses in wheat. Frontiers in plant science. 2016;7: 929.

[93] Mishra RC. and Grover A. Constitutive over-expression of rice ClpD1 protein enhances tolerance to salt and desiccation stresses in transgenic Arabidopsis plants. Plant Science. 2016; 250: 69-78.

[94] Banaei-Asl F, Bandehagh A, Uliaei ED, Farajzadeh D, Sakata K, Mustafa G, Komatsu S. Proteomic analysis of canola root inoculated with bacteria under salt stress. Journal of proteomics. 2015;124:88-111.

[95] Shokri-Gharelo R. and Noparvar PM. Molecular response of canola to salt stress: insights on tolerance mechanisms. PeerJ. 2018;6: e4822.

[96] Zhu JK. Abiotic stress signaling and responses in plants. Cell, 2016; 167(2): 313-324.

[97] Bhattacharya M, Babwah A. and Ferguson S. Small GTP-binding protein-coupled receptors. Portland Press Ltd. 2004.

[98] Ghosh D. and Xu J. Abiotic stress responses in plant roots: a proteomics perspective. Frontiers in plant science, 2014; 5: 6.

[99] Das S, Hussain A, Bock C, Keller WA. and Georges F. Cloning of *Brassica napus* phospholipase C2 (BnPLC2), phosphatidylinositol 3-kinase (BnVPS34) and phosphatidylinositol synthase1 (BnPtdIns S1)—comparative analysis of the effect of abiotic stresses on the expression of phosphatidylinositol signal transduction-related genes in *B. napus*. Planta. 2005; 220(5): 777-784.

[100] Konopka-Postupolska D, Clark G, Goch G, Debski J, Floras K, Cantero A, Fijolek B. The role of annexin 1 in drought stress in Arabidopsis. Plant physiology. 2009; 150(3): 1394-1410.

[101] Zhang H, Yang B, Liu WZ, Li H, Wang L, Wang B, Deng M. Identification and characterization of CBL and CIPK gene families in canola (*Brassica napus* L.). BMC plant biology. 2014a; 14(1): 8.

[102] Lee SC, Lim MH, Kim JA, Lee SI, Kim JS, Jin M, Kwon SJ. Transcriptome analysis in *Brassica rapa* under the abiotic stresses using Brassica 24K oligo microarray. Molecules & Cells (Springer Science & Business Media BV). 2008; 26(6).

[103] Chinnusamy V. and Zhu JK. Epigenetic regulation of stress responses in plants. Current opinion in plant biology. 2009;12(2): 133-139.

[104] Sokol A, Kwiatkowska A, Jerzmanowski A, Prymakowska-Bosak M. Up-regulation of stress-inducible genes in tobacco and Arabidopsis cells in response to abiotic stresses and ABA treatment correlates with dynamic changes in histone H3 and H4 modifications. Planta. 2007;227(1): 245-254.

[105] Labra M, Grassi F, Imazio S, Di Fabio T, Citterio S, Sgorbati S, Agradi E. Genetic and DNA-methylation changes induced by potassium dichromate in *Brassica napus* L. Chemosphere. 2004;54(8): 1049-1058.

[106] Attwood J, Yung R, Richardson B. DNA methylation and the regulation of gene transcription. Cellular and Molecular Life Sciences CMLS. 2002;59(2): 241-257.

[107] Guangyuan L, Xiaoming W, Biyun C, Gao G, Kun X. Evaluation of genetic and epigenetic modification in rapeseed (*Brassica napus*) induced by salt stress. Journal of Integrative Plant Biology. 2007; 49(11): 1599-1607.

[108] Lu X, Shu N, Wang J, Chen X, Wang D, Wang S, Fan W. Genome-wide analysis of salinity-stress induced DNA methylation alterations in cotton (*Gossypium hirsutum* L.) using the Me-DIP sequencing technology. Genetics and molecular research: GMR. 2017; 16(2).

[109] Marconi G, Pace R, Traini A, Raggi L, Lutts S, Chiusano M, Guiducci M. Use of MSAP markers to analyse the effects of salt stress on DNA methylation in rapeseed (*Brassica napus var. oleifera*). PloS one. 2013;8(9): e75597.

[110] Li, C.-J., Cheng, P., Liang, M.-K., Chen, Y.-S., Lu, Q., Wang, J.-Y., Xia, Z.-Y. et al. 2015. MicroRNA-188 regulates age-related switch between osteoblast and adipocyte differentiation. The Journal of clinical investigation, 125(4): 1509-1522.

[111] Hong Y. and Jackson S. Floral induction and flower formation—the role and potential applications of mi RNA s. Plant biotechnology journal. 2015;13(3): 282-292.

[112] Hackenberg M, Gustafson P, Langridge P, Shi BJ. Differential

expression of micro RNA s and other small RNA s in barley between water and drought conditions. Plant biotechnology journal. 2015;13(1): 2-13.

[113] Curaba J, Singh MB, Bhalla PL. miRNAs in the crosstalk between phytohormone signalling pathways. Journal of experimental botany. 2014;65(6): 1425-1438.

[114] Jian J, Tian QY, Hettinghouse A, Zhao S, Liu H, Wei J, Grunig G. Progranulin recruits HSP70 to β-glucocerebrosidase and is therapeutic against Gaucher disease. EBioMedicine. 2016;13: 212-224.

[115] Nakashima K, Takasaki H, Mizoi J, Shinozaki K, Yamaguchi-Shinozaki K. NAC transcription factors in plant abiotic stress responses. Biochimica et Biophysica Acta (BBA)-Gene Regulatory Mechanisms. 2012;1819(2): 97-103.

[116] Wang X, Yu K, Li H, Peng Q, Chen F, Zhang W, Chen S. High-density SNP map construction and QTL identification for the apetalous character in *Brassica napus* L. Frontiers in plant science. 2015;6: 1164.

[117] Zhong H, Guo QQ, Chen L, Ren F, Wang QQ, Zheng Y, Li XB. Two *Brassica napus* genes encoding NAC transcription factors are involved in response to high-salinity stress. Plant cell reports. 2012;31(11): 1991-2003.

[118] Yıldız, M., Akçalı, N. and Terzi, H. 2015. Proteomic and biochemical responses of canola (*Brassica napus* L.) exposed to salinity stress and exogenous lipoic acid. Journal of plant physiology, 179: 90-99.

[119] Deinlein U, Stephan AB, Horie T, Luo W, Xu G, Schroeder JI. Plant salt-tolerance mechanisms. Trends in plant science. 2014;19(6): 371-379.

[120] Gharelo R. and Bandehagh A. Analysis of the Promoter Region of the Gene Encoding Sodium/Hydrogen Exchanger 1 Protein. J Mol Genet Med 2017; 11(312): 1747-0862.1000312.

[121] Gharelo Rs, Bandehagh A, Toorchi M, Farajzadeh D. Canola 2-dimensional proteome profiles under osmotic stress and inoculation with *Pseudomonas fluorescens* FY32. Plant Cell Biotechnol Mol Biol. 2016;17: 257-266.

[122] Jia H, Shao M, He Y, Guan R, Chu P, Jiang H. Proteome dynamics and physiological responses to short-term salt stress in *Brassica napus* leaves. PLoS One. 2015;10(12): e0144808.

[123] Katz A, Waridel P, Shevchenko A, Pick U. Salt-induced changes in the plasma membrane proteome of the halotolerant alga *Dunaliella salina* as revealed by blue native gel electrophoresis and nano-LC-MS/MS analysis. Molecular & Cellular Proteomics. 2007;6(9): 1459-1472.

[124] Wu H. and Zhang Y. Mechanisms and functions of Tet protein-mediated 5-methylcytosine oxidation. Genes & development. 2011; 25(23): 2436-2452.

[125] Chen L, Ren F, Zhou L, Wang QQ, Zhong H. and Li XB. The *Brassica napus* calcineurin B-Like 1/CBL-interacting protein kinase 6 (CBL1/CIPK6) component is involved in the plant response to abiotic stress and ABA signalling. Journal of experimental botany. 2012; 63(17): 6211-6222.

[126] Zhang HX, Hodson JN, Williams JP. and Blumwald E. Engineering salt-tolerant Brassica plants: characterization of yield and seed oil quality in transgenic plants with increased vacuolar sodium accumulation. Proceedings of the National Academy of Sciences. 2001; 98(22): 12832-12836.

[127] Wang W, Zhang H, Wei X, Yang L, Yang B, Zhang L, Li J. Functional characterization of calcium-dependent

protein kinase (CPK) 2 gene from oilseed rape (*Brassica napus* L.) in regulating reactive oxygen species signaling and cell death control. Gene. 2018; 651: 49-56.

[128] Asano T, Hayashi N, Kobayashi M, Aoki N, Miyao A, Mitsuhara I, Ichikawa H. A rice calcium-dependent protein kinase OsCPK12 oppositely modulates salt-stress tolerance and blast disease resistance. The Plant Journal. 2012; 69(1): 26-36.

[129] Ning K, Ng HK, Srihari S, Leong HW, Nesvizhskii AI. Examination of the relationship between essential genes in PPI network and hub proteins in reverse nearest neighbor topology. BMC bioinformatics. 2010;11(1): 505.

[130] Shafeinie A, Mohammadi V, Alizadeh H, Zali AA. Overexpression of Arabidopsis dehydration-responsive element-binding protein 2a confers tolerance to salinity stress to transgenic canola. Pakistan Journal of Biological Sciences. 2014;17(5): 619-629.

[131] Sun Xe, Feng Xx, Li C, Zhang Zp. and Wang Lj. Study on salt tolerance with YHem1 transgenic canola (*Brassica napus*). Physiologia plantarum 2015; 154(2): 223-242.

[132] Osakabe Y, Watanabe T, Sugano SS, Ueta R, Ishihara R, Shinozaki K. and Osakabe K. Optimization of CRISPR/Cas9 genome editing to modify abiotic stress responses in plants. Scientific reports. 2016; 6: 26685.

[133] Haak DC, Fukao T, Grene R, Hua Z, Ivanov R, Perrella G, Li S. Multilevel regulation of abiotic stress responses in plants. Frontiers in plant science. 2017;8: 1564.

[134] Gong F, Hu X, Wang W. Proteomic analysis of crop plants under abiotic stress conditions: where to focus our research? Frontiers in plant science. 2015;6: 418.

Chapter 7

Epidemiology, Genetics and Resistance of *Alternaria* Blight in Oilseed *Brassica*

Subroto Das Jyoti, Naima Sultana, Lutful Hassan and Arif Hasan Khan Robin

Abstract

Alternaria blight is one of the most deadly diseases of oilseed *Brassica*. This recalcitrant disease causes up to 50% yield loss across the globe. The disease is mainly caused by *Alternaria brassicae* and *Alternaria brassicicola*. These pathogens lack sexual stages and survive as conidia or condiospores on the debris of previous crops and susceptible weeds. Developing resistant oilseed *Brassica* cultivars to this disease has become a prime concern for researchers over the years. In absence of resistant oilseed *Brassica* cultivar, identification and introgression of resistance related genes can be a potential source for *Alternaria* blight resistance. As resistance toward *Alternaria* blight is governed by polygenes, intercrossing between the tolerant genotypes and subsequent selection will be the most appropriate way to transfer the quantitative resistance. For that reason, future breeding goal should focus on screening of germplasms for selecting genotypes containing resistance genes and structural features that favors resistance, like thick epicuticular wax, biochemical components such as phenols, phytoalexins and lower soluble sugars, reducing sugars and soluble nitrogen. Selected genotypes should be brought under appropriate breeding programs for attaining *Alternaria* blight resistance.

Keywords: *Alternaria* blight, oilseed *Brassica*, disease resistance, resistance mechanism

1. Introduction

Oilseed crops are one of the crucial pillars of world agriculture, occupying 22% of the world's arable land [1]. Rapeseed-mustard dominates the total oilseed production after soybean globally [1]. *Alternaria* blight is one of the major biotic threats that drastically reduces oilseed production all over the world including Australia, Europe, China and Canada [2]. *Alternaria* blight is a recalcitrant disease caused by the *Alternaria* species primarily *A. brassicae* and *A. brassicicola*, of which *A. brassicae* is the most deadly [3–4]. This disease decreases photosynthetic potential, leads to abnormal growth of the seeds and reduces seed oil content and quality [5]. Disease intensity varies across seasons and regions, and also between crops within an area [6]. Controlling the disease is one of the foremost concerns for researchers for reviving the yield potential of the rapeseed-mustard varieties. Chemical management of this disease is not proposed because maximum foliage coverage by aerial application of fungicides is hard to achieve. Beside this, application of large amounts of chemicals raises environmental concerns. It is crucial to genetically monitor the disease by breeding for resistance [7]. Despite the immense efforts of breeders throughout the world, no resistant genotypes have been found till date. Combining various breeding tools may be fruitful in defining resistant genotypes in these scenarios. The genetic base of the cultivated oilseed *Brassica* is narrow and resistance governing genes are hard to find. *Alternaria* blight resistance is controlled by additives or polygenes and has been identified in some wild species of oilseed *Brassica* [8]. Easy availability of microarray data led researchers to the identification and understanding of the expression patterns of key genes involved in the *Alternaria* resistance. Another reliable form of plant immunity is Nonhost Resistance (NHR) that is successful against all genetic variants of a pathogen [8–11]. The infected plants also show hypersensitive response by producing reactive oxygen species [12]. Improvement of modern genetic transformation methods is helping scientists to incorporate resistant genes from non-host wild cultivars. Tissue culture method is one of the biotechnological tools that are being used to transfer resistance genes from resistant genotypes to the susceptible ones. Resistant genotypes showed higher phenolic content than the susceptible one, whereas the total soluble sugars, lower sugars and soluble nitrogen levels were lower [13–15]. Apart from all of these conventional methods, exploration and utilization of systematically acquired resistance and *de novo* resistance can be an efficient way to induce resistance in oilseed *Brassica* cultivar. Besides, molecular markers associated with resistance genes may contribute to the successful improvement of the resistance breeding process. This chapter discusses *Alternaria* blight disease with respect to its epidemiology, genetics and possible resistance mechanisms involved in *Alternaria* resistance and revisits earlier work done by oilseed *Brassica* breeders to elucidate future strategies for *Alternaria* resistance breeding.

2. Epidemiology

Disease epidemiology provides better understanding of the disease, host and favorable factors that facilitates disease progression. It also creates a better opportunity to control the disease by manipulating different epidemiological factors [16]. Majority of the *Alternaria* species produce asexual spores, as it lacks sexual stage (**Figure 1**; [17]). It survives as conidiospores or conidia under unfavorable conditions [18–19]. It also survives in the susceptible weed and in the infected seeds in temperate regions [20–23]. Although in tropical and subtropical India, the survival of *Alternaria* inoculum in seeds is discarded [24]. At first, symptoms start with black dots. Later, these spots extend and grow into prominent round spots with concentric circles displaying the spot's target board features (**Figure 2**). Many spots coalesce to form large patches which cause the leaves to blight and defoliate [4]. Initially the infection starts from the cotyledonary leaves and forms a basis for the secondary infection. Four hours of leaf wetness is necessary for leaf infection. An increases in leaf wetness duration at 25 °C increases infection and spread of the disease rapidly. Spores attack other parts of the plant upon getting favorable conditions. New lesions arise within four-five days. The pathogen infects the seed by penetrating the pod [25]. The critical factors for spore germination have been reported as darkness or low light intensity (<1000 lux), 25 °C temperature and more than 90% RH in some previous studies [26]. Some studies reported the increase of disease severity with the increase of inoculum concentration [27–29]. The optimal assay temperature of 25 °C and > 90% relative humidity resulted in the highest severity of the disease, regardless of the apparent susceptibility of the

Figure 1.
Cultured spores (a) and conidia of Alternaria brassicae *from the infected field samples (b).*

Figure 2.
Symptoms and different level of severity of Alternaria *blight. Symptoms from 'a' to 'e' show gradually higher severity of infection.*

cultigen [27, 30–36]. Previous studies reported that older leaves are more affected by *Alternaria* than the younger ones [27, 30, 37–40]. Weather characteristics such as maximum temperature 18–27 °C and minimum temperature 8–12 °C facilitates *Alternaria* infection on leaves with an average relative humidity more than 92% while on pods, the infection occurs at temperatures ranging from 20–30 °C [41]. Closer spacing (30 × 15 cm), high nitrogen doses (80 Kg Nha^{-1}) and frequent irrigation rapidly increase severity of disease in rapeseed–mustard [12]. Frequent rains are favorable for the initiation and spread of the disease on the leaves of oilseed *Brassica*. In addition, the rate of infection during the flowering and pod phases is the highest [42].

3. Genetics and genomics of *Alternaria* blight resistance

Identifying resistance mechanisms at the genetic and genomic level has been a prime concern for the researchers over the recent years. Various sources suggest that the resistance against *Alternaria* is polygenic [3, 43–45]. On the contrary, other studies reported that resistance to this disease is mainly controlled by only additive genes or dominant nuclear genes [3, 43–46]. However, Kumar et al. [47] proved that inheritance of *Alternaria* blight resistance is governed by more than one gene and fixable and non-fixable gene effects are vital in the genetic control of *Alternaria* blight resistance. In *Arabidopsis*, six QTLs governing *Alternaria* blight resistance were identified. Among these QTLs, five QTLs were population specific and one was common among all mapping populations. Presence of both common and population specific QTLs indicates that resistance against *Alternaria* blight is quantitative and more than one gene potentially governs the resistance [48].

With the modern development of biotechnology, the discovery of resistance (R) and defense-related genes has opened up new scopes for inducing genetic resistance against different biotic and abiotic stresses [49]. Advances in microarray data processing also ease the process of identifying candidate genes in certain physiological processes. In previous studies, *A. brasscicola* infection contributed to the upregulation of different genes such as *WRKY, peroxidase, p450 oxidases, Chitinase* that modulates defense response in oilseed *Brassica* and *Arabidopsis*. A recent computational study identified vital genes involved in *Alternaria* resistance in *Brassica* by analyzing microarray data of model plant *Arabidopsis thaliana* challenged with *Alternaria* infection [50]. *NHL10, HCHIB* and *XLG2* were identified as major genes and *CZF1, ARF6, WRKY, MP, IAA1, IAA19, AXR3* as candidate genes associated in defense response against *Alternaria* [50]. PR (pathogenesis-related) proteins are a distinct group of molecules which are induced by phytopathogens and signaling molecules linked to defense. They are the vital components of the plant's inherent immune system, particularly systemic acquired resistance (SAR) [51]. Two genes under these proteins namely *Chitinase* and *NPR1* have been characterized in oilseed *Brassica* species. Their high expression level in resistant genotypes compared to the susceptible genotypes suggested that these genes are related to resistance against *Alternaria* blight [52–53]. Another study reported the expression of *PR-3* and *PR-12* only in *Camelina sativa* and *Sinapsis alba* compared with *B. juncea* [54]. This clarifies the involvement of PR proteins in the resistance mechanism of *Alternaria* resistant varieties.

4. Biochemical resistance against *Alternaria*

Biochemical defense is triggered by any stress condition in a plant and is the most important tool of plant defense mechanism. The hypersensitive response is one of the plant's most effective defensive responses against the pathogen [55]. Resistance to *Alternaria* blight in mustard was reported to be linked with the synthesis of phenolic pathway-associated leaf enzymes and higher leaf sugar content [56]. The concentration of phenolic compounds at all stages of plant growth was reported to be high in resistant genotypes compared to susceptible genotypes. Nevertheless, soluble sugars, sugar reduction and soluble nitrogen levels in resistant genotypes were lower [14–15]. Another study reported that, total phenol, total sugar, reducing sugar, o-dihydroxy phenol, chlorophyll content and flavonol contents were higher in resistant genotypes [57]. By activating several defense responses that dissuade the infection process, plants can respond to a pathogen. These include the production of reactive oxygen species (ROS), the accumulation of proteins related to pathogenesis (PR) and phytoalexins and the synthesis of compounds that strengthen the plant cell wall [58]. Moreover the contents of ascorbic acid, total phenol, enzymatic activities of superoxide dismutase and peroxidase, that of cell protecting enzymes such as phenylalanine ammonia lyase and polyphenol oxidases were increased in the resistant genotypes of mustard [59]. *β-Aminobutyric acid* (BABA), a non-protein amino acid has been known to stimulate resistance to a variety of pathogens in a number of plant species [60–61]. Pretreatment of oilseed *Brassica* plants with BABA-mediated resistance to the necrotrophic pathogen *A. brassicae* through enhanced expression of protein genes linked to pathogenesis [62]. The colonization of *A. brassicae* on *Brassica carinata* leaves was substantially inhibited by the foliar application of BABA [63]. A higher and early accumulation of H_2O_2 was observed in resistant *C. sativa* and *S. alba* compared to *B. juncea*. Catalase activity was enhanced in both *C. sativa* and *S. alba*, but the opposite phenomenon was observed in case of *B. juncea* [54].

5. Utilization of non-host resistance

Non-host resistance is one of the most useful approaches for attaining resistance against different plant pathogens. Till date, no resistant cultivar is available in oilseed *Brassica* species. Therefore, utilizing the non-host resistance from wild species can be an efficient breeding tool. Plant pathogens manage to affect different species, but they fail to overcome the non-host resistance [64]. Examples of some non-host plants of *A. brassicae* are chickpea, lentil, wheat, sugarcane, barley, tomato, potato [64]. NHR is multilayered and can be splitted into two main forms: the layer of pre-invasion and the phase of post-invasion [65–67]. Preformed defenses may include structural features like abundance of trichomes and spore germination inhibitory chemical compounds [68–70]. Previous studies reported that spore germination occurs at an equal rate in both host and non-host plants [71]. Despite an accurate germination, pathogens might fail to reach the stomata. Stomata in non-host plants may not be correctly recognized by the pathogen because the topography of the surface may vary significantly from that of the host leaf [64]. Another structural feature that can prevent the entry of *Alternaria* is the epicuticular wax [72–74]. Non-host plants may have higher epicuticular wax than the susceptible host plants [64]. The non-host plant is capable of inducing stomatal closure, preventing pathogens from entering and constructing an inducible chemical barrier that suppresses hyphal production and differentiation by the rapid formation of phytoalexins, antimicrobial compounds [75–77]. In a non-host plant, the dietary deficiency and the presence of antimicrobial compounds in the apoplast can also prevent the production of hyphae into mycelium [71]. The pathogen also generates non-host specific or general toxins that might damage plant cells, leading ultimately to necrosis [78–80]. To avoid this, a non-host plant may recognize these toxins and employ defense mechanisms to detoxify these toxins [81]. In *Arabidopsis* and *S. alba* pathogenesis-related genes *PR-1, PR-2, PR-3* were highly expressed compared to *B. juncea* after *Alternaria* infection [82–86]. Furthermore, these two species showed non-host resistance toward *A. brassicicola* [81, 87]. Chitinase enzymes that hydrolyze the fungal cell wall and release fragments of chitin are actively secreted by these two species [82, 88]. The NHR action includes the stimulation by the plant cell of a signal transduction cascade following the detection of a pathogen, which triggers the activation of protein kinases and mitogen-activated protein kinase (MAPK) members and consequently lead to the activation of defensive genes in non-host plants [89]. The expression of *MAPK* was higher in *S. alba* and downregulated in *B. juncea* suggesting its possible role in *Alternaria* blight resistance.

6. Genetic transformation for *Alternaria* resistance

As the resistance of *Alternaria* has not yet been found, identification of resistance genes in non-host plants and transferring them into oilseed *Brassica* species could be a handy tool for resistance breeding. Introgression of genes under PR-proteins have been found effective in many cases. For instance, transgenic Indian mustard was developed with the *chitinase* gene in which the occurrence of disease symptoms was delayed by a duration of 10–15 days compared to control plants [90]. For enhancing resistance against *A. brassicae*, a PR protein-encoding glucanase was introduced from tomato into Indian mustard plants [91]. Glucanase hydrolyzes a main component of a fungal cell wall called glucan and destroys the invading fungal pathogens. In combating *Alternaria* blight disease, a barley antifungal class II *chitinase* gene and type I ribosome inactivating protein (*RIP*) gene were co-expressed in Indian mustard [92]. Transgenic mustard plants demonstrated a 44% reduction in *A. brassicae* hyphal production relative to the control plants. When transgenic events were sprinkled with fungal spores through greenhouse screening, the late onset of the disease and a lower number of lesions with reduced size distribution were recorded. In addition, *Chitinase* gene was transferred from *Streptomyces griseus* HUT6037 to Indian mustard [93]. A previous study transformed *B. juncea* with the *osmotin* gene and documented resistance to the purified *A. brassicae* toxin in the transformed calli [94]. *B. juncea* was modified to add resistance to *Alternaria* blight and stem rot diseases with the *MSRA1* gene [95]. Bioassays after *Alternaria* infection *in vitro* showed that transgenic *B. juncea* lines inhibited the growth of *Alternaria* hyphae by 44–62% and reduced infection ranging from 69–85%. The *lectin* gene of chickpea was transferred to Indian mustard cv. Varuna to induce resistance against *A. brassicae* in transgenic lines [96]. Another study incorporated *B. juncea* with the gene *MPK3* and examined its role in providing tolerance against *A. brassicae* [97]. In transgenic plants, both ascorbate peroxidase (APX) and guaiacol peroxidase (GP) activity and proline content were higher, leading to the scavenging of ROS in transgenic plants developed as a result of infection with *Alternaria*.

When an *endochitinase* gene '*echh42*' from the *Trichoderma virens*, a fungal species used as a bio-control agent, was introduced to *B. juncea*– the transformed plants showed 7-fold higher endochitinase activity compared to the non-transformed plants based on fluorimetric analysis [98]. These results indicated that the *endochitinase* gene '*ech42*' could be a major gene that may provide resistance to oilseed *Brassica* plants against the *Alternaria* blight. In previous studies, the transgenic broccoli plants also showed expression of *chitinase* gene of *Trichoderma harzianum* [99–101]. Moreover, the synthetic *chitinase* gene (*NIC*) showed broad-spectrum resistance to the transgenic lines of *B. juncea* including *A. brassicicola* [102]. Further research utilizing RT-PCR validated that these *chitinase* genes were induced after wounding and exogenous treatments of jasmonic acid and salicylic acid similar to *Alternaria* infection [103]. A recent review summarized that the chitinases, glucanases or cry proteins provide broad-spectrum resistance against some major diseases including *Alternaria* blight and blackleg [104].

7. *De novo* resistance

It is assumed that the disease can be successfully managed by inducing protection inducers in plants. Some novel fungicides may mimic the action of different plant hormones that activate the plant's internal immune response. Jasmonic acid (JA) mediated defense response to *A. brassicae* fungus can prevent necrotrophic colonization mode. The JA receptor, coronatine insensitive 1 (COI1), is one of the possible targets to activate JA-mediated immunity via JA signal interaction [105]. It is understood that Jasmonates and its functional analogs play a crucial role in systemic defense, likely serving as the initiating signal of acquired systemic resistance [106]. It has been shown that necrotrophic fungal pathogens are the primary activators of JA-dependent defenses via COI_1 receptor activation [107]. A previous study identified some JA mimicking molecules that might be helpful in *de novo* resistance induction [108].

8. Tissue culture techniques in *Alternaria* resistance

Tissue culture is one of the most effective tools of modern biotechnology. Somaclonal variation provides an opportunity to extend the genetic variation of crops, i.e. the variation caused by cell and tissue culture. By applying *in vitro* selection process, the efficiency of selection can be increased [109]. Somatic hybrids were produced through PEG-mediated symmetric and asymmetric protoplast fusion, in which *S. alba*, *B. nigra* and *B. juncea* were found to be the most effective resistance donor to *Alternaria* pathogen [110]. Through protoplast fusion, a previous study developed three hybrids between *B. juncea* and *S. alba* [111]. Among the hybrids, two of the hybrids were symmetric, while the third was asymmetric and had greater similarity to *B. juncea*. *Alternaria* resistant lines were developed through interspecific hybridization between *S. alba* and *B. juncea* [112]. *Alternaria* blight resistance was transferred from *B. tourneforti* to *B. juncea* cv. RH 30 through *in vitro* ovule culture [113]. Intergeneric hybrids of *B. campestris* and *B. spinenscens* were generated through sequential ovary, ovule and embryo culture [114]. The resistance trait was transferred to *B. napus* cv. Brutor from *S. alba* cv. Carine following *in vitro* fertilized ovary culture protocol [115]. *Erucastrum cardaminoides* and *B. oleracea* var. alboglabra were used to develop intergeneric hybrids with *Alternaria* blight resistance following sequential ovary and ovule culture procedures [116]. Previous studies reported transfer of *Alternaria* resistance through somatic hybridization such as, from *S. alba* to *B. napus* [117] and *Moricandida arvensis* to *B. oleracea* [118]. A research group in India transferred *Alternaria* resistance trait to *B. juncea* from *B. carinata* [119]. Disease resistant hybrid plants were produced from the hybridized leaf mesophyll protoplasts of *M. arvensis* and *B. napus* [120]. *B. carinata* was resynthesized by protoplast fusion between *B. nigra* and *B. oleracea* [121]. The hybrids thus obtained were fertile and grew into robust plants. Previous studies conducted hybridization between *S. alba* and *B. oleracea* and between *Camelina sativa* and *B. oleracea* for producing resistant hybrids [122–123]. Another study developed somatic hybrids between *S. alba* and *B. oleracea* by protoplast fusion followed by embryo rescue and managed to recover four highly resistant hybrid progenies after repeated backcrosses [124]. By inducing variations through gamma-irradiated mutagenesis the resistant varieties were obtained in *B. juncea* [125] while another study achieved the similar results by treating the embryos with chemical mutagens [126]. It is plausible to say that proper utilization of tissue culture techniques can be a successful means of incorporating *Alternaria* resistance into oilseed *Brassica* cultivars.

9. Molecular markers and *Alternaria* blight resistance

In any disease resistance breeding program, the primary approach is to quickly screen all the available germplasm including local races, improved variety and exotic genetic stocks. The traditional approach of screening of genotypes can be costly, time and space consuming, laborious, and involves large sample sizes [127]. The limitations of conventional approach can be solved through molecular markers. By utilizing molecular markers, economically important major genes and quantitative trait loci (QTLs) can be identified [128]. Pre-selection using molecular markers can minimize the size of a population and facilitate early detection of desirable genotypes [127]. Various molecular markers are being used nowadays for assessing genetic variability against *Alternaria* blight. For example, internal transcribed spacer regions (ITS), restriction fragment length polymorphism (RFLP), randomly amplified polymorphic DNA (RAPD), amplified fragment length polymorphism (AFLP), inter-simple sequence repeats (ISSRs), microsatellites (SSR), sequence tagged sites (STS), single nucleotide polymorphism (SNPs) etc. The ITS regions are the preserved areas in the fungal genome that are considered as the most common loci to study DNA based mycology at the species level. Berbee and co-workers studied the ITS regions of rDNA to determine the pathogen's phylogeny [129]. RAPD technique was used successfully to examine the genetic differences in *Alternaria* infected species [130–132]. Later on, the assessment of genetic variability in *Alternaria* species has moved to more sensitive techniques such as AFLP [133] and microsatellite markers [134] due to the constraints of reproducibility of RAPD. Simple sequence repeats have been isolated and characterized from *B. napus*, *B. nigra*, and *B. rapa* [135, 136]. Moreover, SSR marker libraries have been developed for *B. rapa* those are being used to produce a genome map for *B. rapa* [137]. Recently, SNP markers have taken the supremacy over SSR as they are unique and plentiful in high and ultra-high-throughput and are able to find polymorphism within a single base pair [138].

10. Conclusions

Alternaria blight is one of the major diseases of oilseed *Brassica* causing enormous yield loss every year. In order to reduce the use of chemical fertilizers and to save the environment, breeding is important to attain resistance against *Alternaria* pathogens. Since the resistance against *Alternaria* blight is governed by additive or polygenes, molecular breeding for resistance could be more effective. All possible sources including wild relatives and non-host plants should be brought under the selection process for identifying ideal resistance donors. QTL mapping and continuous hybridization between resistant genotypes should be performed for better results. Emphasis should be given on functional analysis of PR proteins for engineering *Alternaria* resistance more effectively. In addition, accurate modeling of plant's internal defense responsive pathways can provide new insights on *de novo* and systematically acquired resistance.

Funding

This research is supported by the Ministry of Education, Bangladesh (Project title: Development of oilseed *Brassica* genotypes resistant to *Alternaria* blight through accelerated genetic gain; Project No. 2019/05/MoE).

Conflict of interest

The authors declare no conflict of interest.

Author details

Subroto Das Jyoti, Naima Sultana, Lutful Hassan and Arif Hasan Khan Robin*
Department of Genetics and Plant Breeding, Bangladesh Agricultural University, Mymensingh, Bangladesh

*Address all correspondence to: gpb21bau@bau.edu.bd

References

[1] FAO. World Food and Agriculture - Statistical Yearbook; 2020. Rome. DOI: 10.4060/cb1329en

[2] Zheng X, Koopmann B, Ulber B, von Tiedemann A. A Global Survey on Diseases and Pests in Oilseed Rape—Current Challenges and Innovative Strategies of Control. Frontiers in Agronomy. 2020; 2:1-5. DOI: 10.3389/fagro.2020.590908

[3] Meena PD, Awasthi RP, Chattopadhyay C, Kolte SJ, Kumar A. *Alternaria* blight: a chronic disease in rapeseed-mustard. Journal of Oilseed *Brassica*. 2016; 1:1-1.

[4] Kumar D, Maurya N, Bharati YK, Kumar A, Kumar K, Srivastava K, Chand G, Kushwaha C, Singh SK, Mishra RK, Kumar A. *Alternaria* blight of oilseed *Brassicas*: A comprehensive review. African Journal of Microbiology Research. 2014; 8:2816-29. DOI: 10.5897/AJMR2013.6434

[5] Saharan GS, Mehta DN, Meena PD, Dayal P. *Alternaria* diseases of crucifers: biology, ecology and disease management. Singapore: Springer; 2016.

[6] Meena PD, Rani A, Meena R, Sharma P, Gupta R, Chowdappa P. Aggressiveness, diversity and distribution of *Alternaria brassicae* isolates infecting oilseed *Brassica* in India. African Journal of Microbiology Research. 2012; 6:5249-58. DOI: 10.5897/AJMR12.705

[7] Sharma G, Kumar VD, Haque A, Bhat SR, Prakash S, Chopra VL. *Brassica* coenospecies: a rich reservoir for genetic resistance to leaf spot caused by *Alternaria brassicae*. Euphytica. 2002; 125:411-7. DOI: 10.1023/A:1016050631673

[8] Heath MC. Nonhost resistance and nonspecific plant defenses. Current Opinion in Plant Biology. 2000; 3:315-9. DOI: 10.1016/S1369-5266(00)00087-X

[9] Mysore KS, Ryu CM. Nonhost resistance: how much do we know? Trends in Plant Science. 2004; 9:97-104. DOI: 10.1016/j.tplants.2003.12.005

[10] Senthil-Kumar M, Mysore KS. Nonhost resistance against bacterial pathogens: retrospectives and prospects. Annual Review of Phytopathology. 2013; 51:407-27. DOI: 10.1146/annurev-phyto-082712-102319.

[11] Lee HA, Lee HY, Seo E, Lee J, Kim SB, Oh S, Choi E, Choi E, Lee SE, Choi D. Current understandings of plant nonhost resistance. Molecular Plant-Microbe Interactions. 2017; 30:515. DOI 10.1094/MPMI-10-16-0213-CR.

[12] Lamb C, Dixon RA. The oxidative burst in plant disease resistance. Annual Review of Plant Biology. 1997; 48:251-75. DOI: 10.1146/annurev.arplant.48.1.251

[13] Verma PR, Saharan GS. Monograph of *Alternaria* Diseases of Crucifers. Agriculture and Agri-Food Canada, Saskatoon Research Centre. Technical Bulletin E. 1994; 6:150.

[14] Meena PD, Bala M, Sharma P, Singh D. Interaction and tolerance to *Alternaria brassicae* in Indian mustard (*Brassica juncea*) genotypes. The Indian Society of Oilseeds Research. 2014:136.

[15] Meena PD, Kumar MS, Meena HS, Jambhulkar S, Pathak D, Srivastava S, Gupta R, Singh D, Gurung B, Rai PK. Confirmation of induced tolerance to *Alternaria* blight disease in Indian mustard (*Brassica juncea* L.). Applied Biochemistry and Biotechnology. 2020; 192:965-78. DOI: 10.1007/s12010-020-03362-2

[16] Jones DG. An introduction to plant disease epidemiology. In The

Epidemiology of Plant Diseases. Springer, Dordrecht. 1998; 3-13. DOI: 10.1007/978-94-017-3302-1_1

[17] Mamgain A, Roychowdhury R, Tah J. *Alternaria* pathogenicity and its strategic controls. Research Journal of Biology. 2013; 1:1-9.

[18] Humpherson-Jones FM, Maude RB. Studies on the epidemiology of *Alternaria brassicicola* in *Brassica oleracea* seed production crops. Annals of Applied Biology. 1982; 100:61-71. DOI: 10.1111/j.1744-7348.1982.tb07192.x

[19] Humpherson-Jones FM. Survival of *Alternaria brassicae* and *Alternaria brassicicola* on crop debris of oilseed rape and cabbage. Annals of Applied Biology. 1989; 115:45-50. DOI: 10.1111/j.1744-7348.1989.tb06810.x

[20] Sherf AF, MacNab AA. Vegetable diseases and their control. John Wiley & Sons. New York. 693: 267-269.

[21] Maude RB, Humpherson-Jones FM. Studies on the seed-borne phases of dark leaf spot *Alternaria brassicicola* and grey leaf spot *Alternaria brassicae* of *Brassicas*. Annals of Applied Biology. 1980; 95:311-9. DOI: 10.1111/j.1744-7348.1980.tb04752.x

[22] Shrestha SK, Munk L, Mathur SB. Survival of *Alternaria brassicae* in seeds and crop debris of rapeseed and mustard in Nepal. Seed Science and Technology. 2003; 31:103-9. DOI: 10.15258/sst.2003.31.1.11

[23] Ahmad MS, Sinha KK. Mycotoxin contamination of mustard seeds during storage. Indian Phytopathology. 2002; 55:299-302.

[24] Mehta N, Sangwan MS, Srivastava MP, Kumar R. Survival of *Alternaria brassicae* causing *Alternaria* blight rapeseed-mustard. Journal of Mycology and Plant Pathology. 2002; 32: 64-67.

[25] Saharan GS, Mehta N, Sangwan MS, editors. Diseases of oilseed crops. Indus Publishing. 2005.

[26] Mehta N. Epidemiology and forecasting for the management of rapeseed-mustard diseases. Journal of Mycology and Plant Pathology. 2014; 44:131.

[27] Hong CX, Fitt BD. Effects of inoculum concentration, leaf age and wetness period on the development of dark leaf and pod spot (*Alternaria brassicae*) on oilseed rape (*Brassica napus*). Annals of Applied Biology. 1995; 127:283-95. DOI: 10.1111/j.1744-7348.1995.tb06673.x

[28] Meena PD, Jambhulkar SJ, Gupta R, Meena HS, Singh D. Rapid screening technique for *Alternaria* blight resistance in Indian mustard (*Brassica juncea* L.) using cotyledonary leaf method. Journal of Plant Pathology. 2016: 463-9.

[29] Nowakowska M, Wrzesińska M, Kamiński P, Szczechura W, Lichocka M, Tartanus M, Kozik EU, Nowicki M. *Alternaria brassicicola*–Brassicaceae pathosystem: insights into the infection process and resistance mechanisms under optimized artificial bio-assay. European Journal of Plant Pathology. 2019; 153:131-51. DOI: 10.1007/s10658-018-1548-y

[30] Doullah MA, Meah MB, Okazaki K. Development of an effective screening method for partial resistance to *Alternaria brassicicola* (dark leaf spot) in *Brassica rapa*. European Journal of Plant Pathology. 2006; 116:33-43. DOI: 10.1007/s10658-006-9035-2

[31] Gupta P, Ravi I, Sharma V. Induction of β-1, 3-glucanase and chitinase activity in the defense response of *Eruca sativa* plants against the fungal pathogen *Alternaria brassicicola*. Journal of Plant Interactions. 2013; 8:155-61. DOI: 10.1080/17429145.2012.679705

[32] Mazumder M, Das S, Saha U, Chatterjee M, Bannerjee K, Basu D. Salicylic acid-mediated establishment of the compatibility between *Alternaria brassicicola* and *Brassica juncea* is mitigated by abscisic acid in *Sinapis alba*. Plant Physiology and Biochemistry. 2013; 70:43-51. DOI: 10.1016/j.plaphy.2013.04.025

[33] Rashid MM, Hossain I, K. M. Khalequzzaman KM. Effect of weather factors on inoculum density and leaf spot development in radish seed crop infected with *Alternaria brassicae*. Bulletin of the Institute of Tropical Agriculture, Kyushu University. 2011; 34:43-7. DOI: 10.11189/bita.34.43

[34] Sharma G, Kumar VD, Haque A, Bhat SR, Prakash S, Chopra VL. *Brassica* coenospecies: a rich reservoir for genetic resistance to leaf spot caused by *Alternaria brassicae*. Euphytica. 2002; 125:411-7.

[35] Su'udi M, Kim MG, Park SR, Hwang DJ, Bae SC, Ahn IP. *Arabidopsis* cell death in compatible and incompatible interactions with *Alternaria brassicicola*. Molecules and Cells. 2011; 31:593-601.

[36] Zală CR, Cristea ST, Gruia LI, Manole SA. Research on the biology of the *Alternaria brassicae* fungus isolated from mustard. In 3rd International Conference on Energy and Environment Technologies and Equipment. Brasov, Romania. p 2014 (pp. 161-164).

[37] Allen SJ, Brown JF, Kochman JK. Effects of leaf age, host growth stage, leaf injury, and pollen on the infection of sunflower by *Alternaria helianthi*. Phytopathology. 1983; 73:896-8.

[38] Deep SW, Sharma PR. Host age as predisposing factor for incidence of black leaf spot of cauliflower caused by *Alternaria brassicae* and *Alternaria brassicicola*. Indian Phytopathology. 2012; 65:71-5.

[39] Rotem J. The biotic and physiological components of pathogenesis. The Genus *Alternaria*; Biology, Epidemiology, and Pathogenicity, The American Phytopathological Society, St. Paul, MN. 1998:95-118.

[40] Saharan GS, Mehta N. Fungal diseases of rapeseed-mustard. Diseases of Field Crops. 2002:193-228.

[41] Aneja JK, Agnihotri A. *Alternaria* blight of oilseed *Brassicas*: epidemiology and disease control strategies with special reference to use of biotechnological apporoaches for attaining host resistance. Journal of Oilseed Brassica. 2016; 1:1-10.

[42] Prasad RA, Saxena DE, Chandra S. Yield losses by *Alternaria* blight in promising genotypes of Indian mustard. Indian Phytopathology. 2003; 56(2):205-6.

[43] Tripathi NN, Kaushik CD, Yadava TP, Yadav AK. *Alternaria* leaf spot resistance in raya. Haryana Agricultural University Journal of Research. 1980; 10:166-8.

[44] Zhang FL, Xu JB, Takahata Y. Inheritance of resistance to black leaf spot (*Alternaria brassicae*) in Chinese cabbage. Cruciferae Newsletter (United Kingdom). 1996.

[45] Krishnia SK, Saharan GS, Singh D. Genetics of *Alternaria* blight resistance in inter and intraspecific crosses of *Brassica juncea* and *Brassica carinata*. Annals of Biology. 2000; 16:211-6.

[46] Panja BN, De DK: Combining ability and heterosis for resistant to leaf blight disease in Indian mustard [*Brassica juncea* (L.) Czern & Coss]. Indian Journal of Genetics. 2005; 65: 55-6.

[47] Kumar V, Kumari V, Meena PD, Kumar A. Role of inter genic

interactions in inheritance of *Alternaria* leaf blight resistance in inter and intraspecific crosses of *Brassica carinata* and *Brassica juncea*. Journal of Oilseed Brassica. 2020; 11:126-8.

[48] Rajarammohan S, Kumar A, Gupta V, Pental D, Pradhan AK, Kaur J. Genetic architecture of resistance to *Alternaria brassicae* in *Arabidopsis thaliana*: QTL mapping reveals two major resistance-conferring loci. Frontiers in Plant Science. 2017; 8:260. DOI: 10.3389/fpls.2017.00260

[49] Pandolfi V, Ribamar Costa Ferreira Neto J, Daniel da Silva M, Lindinalva Barbosa Amorim L, Carolina Wanderley-Nogueira A, Lane de Oliveira Silva R, Akio Kido E, Crovella S, Maria Benko Iseppon A. Resistance (R) genes: applications and prospects for plant biotechnology and breeding. Current Protein and Peptide Science. 2017; 18:323-34.

[50] Pathak RK, Baunthiyal M, Pandey D, Kumar A. Computational analysis of microarray data of *Arabidopsis thaliana* challenged with *Alternaria brassicicola* for identification of key genes in *Brassica*. Journal of Genetic Engineering and Biotechnology. 2020; 18:1-20. DOI: 10.1186/s43141-020-00032-y

[51] Ali S, Ganai BA, Kamili AN, Bhat AA, Mir ZA, Bhat JA, Tyagi A, Islam ST, Mushtaq M, Yadav P, Rawat S. Pathogenesis-related proteins and peptides as promising tools for engineering plants with multiple stress tolerance. Microbiological Research. 2018; 212:29-37. DOI: 10.1016/j.micres.2018.04.008

[52] Ali S, Mir ZA, Tyagi A, Mehari H, Meena RP, Bhat JA, Yadav P, Papalou P, Rawat S, Grover A. Overexpression of *NPR1* in *Brassica juncea* confers broad spectrum resistance to fungal pathogens. Frontiers in Plant Science. 2017; 8:1693. DOI: 10.3389/fpls.2017.01693

[53] Mir ZA, Ali S, Shivaraj SM, Bhat JA, Singh A, Yadav P, Rawat S, Paplao PK, Grover A. Genome-wide identification and characterization of *Chitinase* gene family in *Brassica juncea* and *Camelina sativa* in response to *Alternaria brassicae*. Genomics. 2020; 112:749-63. DOI: 10.1016/j.ygeno.2019.05.011

[54] Yadav P, Mir ZA, Ali S, Papolu PK, Grover A. A combined transcriptional, biochemical and histopathological study unravels the complexity of *Alternaria* resistance and susceptibility in *Brassica* coenospecies. Fungal Biology. 2020; 124:44-53.

[55] Goodman RN, Novacky AJ. The hypersensitive reaction in plants to pathogens: a resistance phenomenon. American Phytopathological Society (APS); 1994.

[56] Singh DN, Singh NK, Srivastava S. Biochemical and morphological characters in relation to *Alternaria* blight resistance in rapeseed-mustard. Annual Agriculture Research. 1999; 20:472-7.

[57] Mathpal P, Punetha H, Tewari AK, Agrawal S. Biochemical defense mechanism in rapeseed-mustard genotypes against *Alternaria* blight disease. Journal of Oilseed *Brassica*. 2016; 1:87-94.

[58] Hammond-Kosack KE, Jones JD. Resistance gene-dependent plant defense responses. The Plant Cell. 1996; 8:1773. DOI: 10.1105/tpc.8.10.1773

[59] Mallick SA, Kumari P, Gupta M, Gupta S. Effect of *Alternaria* blight infection on biochemical parameters, quantity and quality of oil of mustard genotypes. Indian Journal of Plant Physiology. 2015; 20:310-316.

[60] Jakab G, Cottier V, Toquin V, Rigoli G, Zimmerli L, Métraux JP, Mauch-Mani B. β-Aminobutyric acid-induced resistance in plants. European

Journal of Plant Pathology. 2001; 107:29-37. DOI: 10.1023/A:1008730721037

[61] Cohen YR. β-aminobutyric acid-induced resistance against plant pathogens. Plant Disease. 2002; 86:448-57. DOI: 10.1094/PDIS.2002.86.5.448

[62] Kamble A, Bhargava S. β-Aminobutyric Acid-induced resistance in *Brassica juncea* against the necrotrophic pathogen *Alternaria brassicae*. Journal of Phytopathology. 2007; 155:152-8. DOI: 10.1111/j.1439-0434.2007.01209.x

[63] Chavan V, Bhargava S, Kamble A. Temporal modulation of oxidant and antioxidative responses in *Brassica carinata* during β-aminobutyric acid-induced resistance against *Alternaria brassicae*. Physiological and Molecular Plant Pathology. 2013; 83:35-9. DOI: 10.1016/j.pmpp.2013.03.002

[64] Fatima U, Bhorali P, Borah S, Senthil-Kumar M. Perspectives on the utilization of resistance mechanisms from host and nonhost plants for durable protection of *Brassica* crops against *Alternaria* blight. PeerJ. 2019; 7:e7486. DOI: 10.7717/peerj.7486

[65] Nurnberger T, Lipka V. Non-host resistance in plants: new insights in an old phenomenon. Molecular Plant Pathology. 2005; 6:335-345. DOI 10.1111/j.1364-3703.2005.00279.x.

[66] Lee HA, Lee HY, Seo E, Lee J, Kim SB, Oh S, Choi E, Choi E, Lee SE, Choi D. Current understandings of plant nonhost resistance. Molecular Plant-Microbe Interactions. 2017; 30:5-15. DOI 10.1094/MPMI-10-16-0213-CR.

[67] Fonseca JP, Mysore KS. Genes involved in nonhost disease resistance as a key to engineer durable resistance in crops. Plant Science. 2018; 279:108-116.

[68] Heath MC. Evolution of plant resistance and susceptibility to fungal parasites. In: Carroll GC, Tudzynski P, editors. The Mycota V, part B, plant relationships. Berlin: Springer. 1997; 257-276.

[69] Heath MC. Signalling between pathogenic rust fungi and resistant or susceptible host plants. Annuals of Botany. 1997; 80:713-720. DOI 10.1006/anbo.1997.0507.

[70] Morrisey JP, Osbourn AE. Fungal resistance to plant antibiotics as a mechanism of pathogenesis. Microbiology and Molecular Biology Reviews. 1999; 63:708-724. DOI:10.1128/MMBR.63.3.708-724.1999

[71] McRoberts N, Lennard JH. Pathogen behaviour and plant cell reactions in interactions between *Alternaria* species and leaves of host and nonhost plants. Plant Pathology. 1996; 45:742-752. DOI: 10.1046/j.1365-3059.1996.d01-4.x

[72] Tewari JP, Skoropad WP. Relationship between epicuticular wax and blackspot caused by *Alternaria brassicae* in three lines of rapeseed. Canadian Journal of Plant Science. 1976; 56:781-785. DOI 10.4141/cjps76-127.

[73] Skoropad WP, Tewari JP. Field evaluation of the epicuticular wax in rapeseed and mustard in resistance to *Alternaria brassicae*. Canadian Journal of Plant Science. 1997; 57:1001-1003. DOI 10.4141/cjps77-146.

[74] Conn KL. Leaf epicuticular wax of canola: ultrastructure, chemistry and interaction with *Alternaria brassicae*. M. S.C. thesis, University of Alberta, Edmonton.1986; 159.

[75] Grayer RJ, Harborne JJ. A survey of antifungal compounds from higher plants. Phytochemistry. 1994; 37:19-42. DOI 10.1016/0031-9422(94)85005-4.

[76] Mert-Turk F. Phytoalexins: defence or just a response to stress? Journal of Cell and Molecular Biology. 2002; 1:1-6.

[77] Iriti M, Faoro F. Review on innate and specific immunity in plants and animals. Mycopathology. 2007; 164:57-64. DOI 10.1007/s11046-007-9026-7.

[78] Buchwaldt L, Green H. Phytotoxicity of destruxin B and its possible role in the pathogenesis of *Alternaria brassicae*. Plant Pathology. 1992; 41:55-63. DOI 10.1111/j.1365-3059.1992.tb02316.x.

[79] Bains PS, Tewari JP, Ayer WA. A note on phytotoxicity of homodestruxin B–a compound produced by *Alternaria brassicae*. Phytoprotection. 1993; 74:157-160.

[80] Parada RY, Oka K, Yamagishi D, Kodama M, Otani H. Destruxin B produced by *Alternaria brassicae* does not induce accessibility of host plants to fungal invasion. Physiology and Molecular Plant Pathology. 2007; 71:48-54. DOI 10.1016/j.pmpp.2007.10.003.

[81] Pedras MSC, Zaharia IL, Gai Y, Zhou Y, Ward DE. In planta sequential hydroxylation and glycosylation of a fungal phytotoxin: avoiding cell death and overcoming the fungal invader. Proceedings of the National Academy of Sciences of the United States of America. 2001; 98:747-752. DOI 10.1073/pnas.98.2.747.

[82] Narusaka Y, Narusaka M, Seki M, Ishida J, Nakashima M, Kamiya A, Enju A, Sakurai T, Satoh M, Kobayashi M. The cDNA microarray analysis using an *Arabidopsis* pad3 mutant reveals the expression profiles and classification of genes induced by *Alternaria brassicicola* attack. Plant Cell and Physiology. 2003; 44:377-387. DOI 10.1093/pcp/pcg050.

[83] Van Wees SC, Chang HS, Zhu T, Glazebrook J. Characterization of the early response of Arabidopsis to *Alternaria brassicicola* infection using expression profiling. Plant Physiology. 2003: 132:606-617. DOI 10.1104/pp.103.022186.

[84] Ghose K, Dey S, Barton H, Loake GJ, Basu D. Differential profiling of selected defence-related genes induced on challenge with *Alternaria brassicicola* in resistant white mustard and their comparative expression pattern in susceptible India mustard. Molecular Plant Pathology. 2008; 9:763-775. DOI 10.1111/j.1364-3703.2008.00497.x.

[85] Nayanakantha NMC, Rawat S, Ali S, Grover A. Differential expression of defense-related genes in *Sinapis alba* and *Brassica juncea* upon the infection of *Alternaria brassicae*. Tropical and Agricultural Research. 2016; 27:123-136.

[86] Mandal S, Rajarammohan S, Kaur J. *Alternaria brassicae* interactions with the model Brassicaceae member *Arabidopsis thaliana* closely resembles those with Mustard (*Brassica juncea*). Physiology and Molecular Biology of Plants. 2018; 24: 51-59. DOI: 10.1007/s12298-017-0486-z

[87] Thomma BP, Eggermont K, Tierens KF, Broekaert WF. Requirement of functional ethylene-insensitive 2 gene for efficient resistance of Arabidopsis to infection by *Botrytis cinerea*. Plant Physiology. 1999; 121:1093-110.1 DOI 10.1104/pp.121.4.1093.

[88] Chatterjee M, Mazumder M, Basu D. Functional analysis of the promoter of a glycosyl hydrolase gene induced in resistant *Sinapis alba* by *Alternaria brassicicola*. Phytopathology. 2013; 103:841-850. DOI 10.1094/PHYTO-11-12-0303-R.

[89] Lawrence CB, Mitchell TK, Craven KD, Cho YR, Cramer RA, Kim KH. At death's door: *Alternaria* pathogenicity mechanisms. Plant Pathology Journal. 2008; 24:101-111. DOI: 10.5423/PPJ.2008.24.2.101.

[90] Mondal KK, Chatterjee SC, Viswakarma N, Bhattacharya RC, Grover A. Chitinase-mediated inhibitory activity of *Brassica* transgenic on growth of *Alternaria brassicae*. Current Microbiology. 2003; 47:0171-3. DOI: 10.1007/s00284-002-3980-6

[91] Mondal KK, Bhattacharya RC, Koundal KR, Chatterjee SC. Transgenic Indian mustard (*Brassica juncea*) expressing tomato glucanase leads to arrested growth of *Alternaria brassicae*. Plant Cell Reports. 2007; 26:247-52. DOI: 10.1007/s00299-006-0241-3

[92] Chhikara S, Chaudhury D, Dhankher OP, Jaiwal PK. Combined expression of a barley class II chitinase and type I ribosome inactivating protein in transgenic *Brassica juncea* provides protection against *Alternaria brassicae*. Plant Cell, Tissue and Organ Culture (PCTOC). 2012; 108:83-9. DOI: 10.1007/s11240-011-0015-7

[93] Ahmad B, Ambreen Ms, Khan I. *Agrobacterium*-mediated transformation of *Brassica juncea* (L.) Czern. with *chitinase* gene conferring resistance against fungal infections. Pakistan Journal of Botany. 2015; 47:211-6.

[94] Taj G, Kumar A, Bansal KC, Garg GK. Introgression of *osmotin* gene for creation of resistance against *Alternaira* blight by perturbation of cell cycle machinery. Indian Journal of Biotechnology. 2004, 3: 291-298.

[95] Rustagi A, Kumar D, Shekhar S, Yusuf MA, Misra S, Sarin NB. Transgenic *Brassica juncea* plants expressing *MsrA1*, a synthetic cationic antimicrobial peptide, exhibit resistance to fungal phytopathogens. Molecular Biotechnology. 2014; 56:535-45. DOI: 10.1007%2Fs12033-013-9727-8

[96] Kumar D, Shekhar S, Bisht S, Kumar V, Varma A, Kumar M. Ectopic overexpression of lectin in transgenic *Brassica juncea* plants exhibit resistance to fungal phytopathogen and showed alleviation to salt and drought stress. Journal of Bioengineering and Biomedical Science. 2015; 5:147. DOI: 10.4172/ 2155-9538-1000147

[97] Tasleem M, Baunthiyal M, Kumar A, Taj G. Determination of antioxidant activity in overexpressed *MPK3* transgenic *Brassica juncea* for induction of defense against *Alternaria* blight disease. Journal of Pharmacognosy and Phytochemistry. 2017; 6:2579-82.

[98] Kamble S, Mukherjee PK, Eapen S. Expression of an endochitinase gene from *Trichoderma virens* confers enhanced tolerance to *Alternaria* blight in transgenic *Brassica juncea* (L.) czern and coss lines. Physiology and Molecular Biology of Plants. 2016; 22: 69-76.

[99] Mora A, Earle E. Combination of *Trichoderma harzianum* endochitinase and a membrane-affecting fungicide on control of *Alternaria* leaf spot in transgenic broccoli plants. Applied Microbiology and Biotechnology. 2001; 55: 306-310.

[100] Mora AA, Earle ED. Resistance to *Alternaria brassicicola* in transgenic broccoli expressing a *Trichoderma harzianum* endochitinase gene. Molecular Breeding. 2001; 8:1-9.

[101] Singh A, Shukla N, Kabadwal BC, Tewari AK, Kumar J. Review on plant-*Trichoderma*-pathogen interaction. International Journal of Current Microbiology and Applied Sciences. 2018; 7: 2382-2397.

[102] Munir I, Hussan W, Kazi M, Farhatullah A, Iqbal A, Munir R. Production of transgenic *Brassica juncea* with the synthetic chitinase gene (NIC) Conferring resistance to *Alternaria brassicicola*. Pakistan Journal of Botany. 2016; 48: 2063-2070.

[103] Rawat S, Ali S, Mittra B, Grover A. Expression analysis of chitinase upon

challenge inoculation to Alternaria wounding and defense inducers in *Brassica juncea*. Biotechnology Reports. 2017; 13: 72-79.

[104] Poveda J, Francisco M, Cartea ME, Velasco P. Development of Transgenic *Brassica* Crops against Biotic Stresses Caused by Pathogens and Arthropod Pests. Plants. 2020; 9: 1664.

[105] Pathak RK, Baunthiyal M, Pandey N, Pandey D, Kumar A. Modeling of the jasmonate signaling pathway in *Arabidopsis thaliana* with respect to pathophysiology of *Alternaria* blight in *Brassica*. Scientific Reports. 2017; 7:1-2.

[106] Truman W, Bennett MH, Kubigsteltig I, Turnbull C, Grant M. *Arabidopsis* systemic immunity uses conserved defense signaling pathways and is mediated by jasmonates. Proceedings of the national academy of sciences. 2007; 104:1075-80. DOI: 10.1073/pnas.0605423104

[107] Antico CJ, Colon C, Banks T, Ramonell KM. Insights into the role of jasmonic acid-mediated defenses against necrotrophic and biotrophic fungal pathogens. Frontiers in Biology. 2012; 7:48-56. DOI: 10.1007%2Fs11515-011-1171-1

[108] Pathak RK, Baunthiyal M, Shukla R, Pandey D, Taj G, Kumar A. *In Silico* Identification of mimicking molecules as defense inducers triggering jasmonic acid mediated immunity against *Alternaria* blight disease in *Brassica* species. Frontiers in Plant Science. 2017; 8:609. DOI: 10.3389/fpls.2017.00609

[109] Van den Bulk RW. Application of cell and tissue culture and in vitro selection for disease resistance breeding—a review. Euphytica. 1991; 56:269-85. DOI: 10.1007/BF00042373

[110] Scholze P, Krämer R, Ryschka U, Klocke E, Schumann G. Somatic hybrids of vegetable *Brassicas* as source for new resistances to fungal and virus diseases. Euphytica. 2010; 176:1-4. DOI: 10.1007/s10681-010-0205-0

[111] Kumari P, Bisht DS, Bhat SR. Stable, fertile somatic hybrids between *Sinapis alba* and *Brassica juncea* show resistance to *Alternaria brassicae* and heat stress. Plant Cell, Tissue and Organ Culture (PCTOC). 2018; 133:77-86. DOI: 10.1007/s11240-017-1362-9

[112] Yadav RC, Singh D, Rathi AR, Arya R, Singh R, Yadav NR, Singh A. Development of *Alternaria* blight resistant lines through interspecific hybridization between Indian mustard (*Brassica juncea* L.) and white mustard (*Brassica alba*) through embryo rescue. Journal of Oilseed *Brassica*. 2018:104-13.

[113] Yadav RC, Sareen PK, Chowdhury JB. Interspecific hybridization in *Brassica juncea* x *Brassica tournefortii* using ovary culture. Cruciferae Newsletter. 1991;84:14-5.

[114] Agnihotri A, Lakshmikumaran MS, Jagannathan V, Shivanna KR. Wide hybridization for improvement in cultivated *Brassicas*. In: International Symposium on Plant Biotechnology and its Contribution to Plant Development, Multiplication and Improvement 289. 1991 (pp. 213-214).

[115] Chèvre AM, Eber F, Margale E, Kerlan MC, Primard C, Vedel F, Delseny M, Pelletier G. Comparison of somatic and sexual *Brassica napus–Sinapis alba* hybrids and their progeny by cytogenetic studies and molecular characterization. Genome. 1994;37: 367-74. DOI: 10.1139/g94-052

[116] Mohanty A, Chrungu B, Verma N, Shivanna KR. Broadening the genetic base of crop *Brassicas* by production of new intergeneric hybrid. Czech Journal of Genetics and Plant Breeding. 2009; 45: 117-22. DOI: 10.17221/35/2009-CJGPB

[117] Primard C, Vedel F, Mathieu C, Pelletier G, Chèvre AM. Interspecific somatic hybridization between *Brassica napus* and *Brassica hirta* (*Sinapis alba L.*). Theoretical and Applied Genetics. 1988; 75:546-52. DOI: 10.1007%2FBF00289119

[118] Toriyama K, Hinata K, Kameya T. Production of somatic hybrid plants,'Brassicomoricandia', through protoplast fusion between *Moricandia arvensis* and *Brassica oleracea*. Plant Science. 1987; 48:123-8. DOI: 10.1016/0168-9452(87)90139-7

[119] Sharma TR, Singh BM. Transfer of resistance to *Alternaria brassicae* in *Brassica juncea* through interspecific hybridization among *Brassicas*. Journal of Genetics & Breeding. 1992; 46:373-8.

[120] O'Neill CM, Murata T, Morgan CL, Mathias RJ. Expression of the C 3-C 4 intermediate character in somatic hybrids between *Brassica napus* and the C 3-C 4 species *Moricandia arvensis*. Theoretical and Applied Genetics. 1996; 93:1234-41. DOI: 10.1007%2FBF00223455

[121] Jourdan P, Salazar E. *Brassica carinata* resynthesized by protoplast fusion. Theoretical and Applied Genetics. 1993; 86:567-72. DOI: 10.1007%252FBF00838710

[122] Hansen LN, Earle ED. Somatic hybrids between *Brassica oleracea* L. and *Sinapis alba* L. with resistance to *Alternaria brassicae* (Berk.) Sacc. Theoretical and Applied Genetics. 1997; 94:1078-85. DOI: 10.1007%2Fs001220050518

[123] Hansen LN. Intertribal somatic hybridization between rapid cycling *Brassica oleracea* L. and *Camelina sativa* (L.) Crantz. Euphytica. 1998; 104:173-9. DOI: 10.1023/A: 1018604919488

[124] Sigareva MA, Earle ED. Production and charactyerization of somatic hybrids between *Camilina sativa* and rapid-cycling *B. oleracea*. Theoretical and Applied Genetics. 1999; 98:164-70. DOI: 10.1007/s001220051053

[125] Verma, VD, Rai, B. Note on induced mutagenesis for spotting out the sources of resistance to *Alternaria* leaf spot in Indian mustard. Indian Journal of Agricultural Sciences. 1980; 50: 278.

[126] Agnihotri A, Gupta K, Prem D, Sarkar G, Mehra VS, Zargar SM. Genetic enhancement in rapeseed-mustard for quality and disease resistance through in vitro techniques. Proceedings of 16th Australian research assembly on *Brassicas*, Ballarat, Australia. 2009; 28.

[127] Ghosh S, Mazumder M, Mondal B, Mukherjee A, De A, Bose R, Das S, Bhattacharyya S, Basu D. Morphological and SSR marker-based genetic diversity analysis of Indian mustard (*Brassica juncea* L.) differing in *Alternaria brassicicola* tolerance. Euphytica. 2019; 215: 206. DOI: 10.1007/s10681-019-2523-1.

[128] Suwabe K, Iketani H, Nunome T, Kage T, Hirai M. Isolation and characterization of microsatellites in *Brassica rapa* L. Theoretical and Applied Genetics. 2002; 104:1092-8. DOI: 10.1007/s00122-002-0875-7.

[129] Berbee ML, Payne BP, Zhang G, Roberts RG, Turgeon BG. Shared ITS DNA substitutions in isolates of opposite mating type reveal a recombining history for three presumed asexual species in the filamentous ascomycete genus *Alternaria*. Mycological research. 2003; 107: 169-82. DOI: 10.1017/S0953756203007263.

[130] Cooke DE, Forster JW, Jenkins PD, Jones DG, Lewis DM. Analysis of intraspecific and interspecific variation in the genus *Alternaria* by the use of RAPD-PCR. Annals of Applied Biology.

1998; 132:197-209. DOI: 10.1111/j.1744 7348.1998.tb05197.x

[131] Sharma TR, Tewari JP. Detection of genetic variation in *Alternaria brassicae* by RAPD fingerprints. Journal of Plant Biochemistry and Biotechnology. 1995; 4: 105-7. DOI: 10.1007/BF03262963

[132] Gherbawy YA. Genetic variation among isolates of *Alternaria* spp. from select Egyptian crops. Archives of Phytopathology and Plant Protection. 2005; 38: 77-89. DOI: 10.1080/03235400400027390

[133] Bock CH, Thrall PH, Brubaker CL, Burdon JJ. Detection of genetic variation in *Alternaria brassicicola* using AFLP fingerprinting. Mycological Research. 2002; 106: 428-34. DOI: 10.1017/S0953756202005762

[134] Avenot H, Dongo A, Bataillé-Simoneau N, Iacomi-Vasilescu B, Hamon B, Peltier D, Simoneau P. Isolation of 12 polymorphic microsatellite loci in the phytopathogenic fungus *Alternaria brassicicola*. Molecular Ecology Notes. 2005; 5: 948-50. DOI: 10.1111/j.1471-8286.2005.01125.x

[135] Lowe AJ, Jones AE, Raybould AF, Trick M, Moule CL, Edwards KJ. Transferability and genome specificity of a new set of microsatellite primers among *Brassica* species of the U triangle. Molecular Ecology Notes. 2002; 2:7-11. DOI: 10.1046/j.1471-8286.2002.00126.x

[136] Piquemal J, Cinquin E, Couton F, Rondeau C, Seignoret E, Perret D, Villeger MJ, Vincourt P, Blanchard P. Construction of an oilseed rape (*Brassica napus* L.) genetic map with SSR markers. Theoretical and Applied Genetics. 2005;111: 1514-23. DOI: 10.1007/s00122-005-0080-6

[137] Ramchiary N, Nguyen VD, Li XI, Hong CP, Dhandapani VI, Choi SR, Yu GE, Piao ZY, Lim YP. Genic microsatellite markers in *Brassica rapa*: development, characterization, mapping, and their utility in other cultivated and wild *Brassica* relatives. DNA research. 2011; 18: 305-20. DOI: 10.1093/dnares/dsr017

[138] Mammadov J, Aggarwal R, Buyyarapu R, Kumpatla S. SNP markers and their impact on plant breeding. International Journal of Plant Genomics. 2012; 2012.

Chapter 8

Breeding for Disease Resistance in Brassica Vegetables Using DNA Marker Selection

Mst Arjina Akter, Hasan Mehraj, Takeru Itabashi, Tomoe Shindo, Masaaki Osaka, Ayasha Akter, Naomi Miyaji, Naoki Chiba, Junji Miyazaki and Ryo Fujimoto

Abstract

The Brassica genus comprises of agro-economically important vegetables. Disease causes great yield loss of Brassica vegetables worldwide. Different traditional methods such as crop rotation and chemical control have limited effect on different diseases of Brassica vegetables and cannot completely eradicate the pathogens by these methods. Development of disease resistant cultivars is one of the most effective, ecofriendly, and cheapest measure to control Brassica diseases. With the development of genomics, molecular biology techniques, and biological methods, it is possible to discover and introduce resistance (*R*) genes to efficiently control the plant diseases caused by pathogens. Some *R* genes of major diseases such as Fusarium wilt and clubroot in Brassica vegetables have been already identified. Therefore, we will focus to review the Fusarium wilt and clubroot resistance in Brassica vegetables and the methodologies for identification, mapping, and pyramiding of *R* genes/quantitative trait loci (QTLs) to develop disease resistant cultivars. These techniques will be helpful for sustainable crop production and to maintain global food security and contribute to ensure protection of food supply in the Asian country as well as throughout the world.

Keywords: *R* gene, marker assisted selection, Fusarium wilt, clubroot, Brassica

1. Introduction

Brassica is a commercially important genus that contains vegetables, oilseeds, condiments, and fodder crops, and they provide nutrition and health-promoting substances to humans worldwide [1]. The commercially important vegetables such as Chinese cabbage (var. *pekinensis*), pak choi (var. *chinensis*), and turnip (var. *rapa*) are involved in *Brassica rapa* L., and cabbage (var. *capitata*), broccoli (var. *italica*), and cauliflower (var. *botrytis*) are involved in *Brassica oleracea* L. [1].

Production of Brassica vegetables constantly threatened by emerging viral, bacterial, and fungal diseases, whose incidence has increased in recent years [2, 3]. The major diseases of Brassica vegetables are Black rot, clubroot, Downy mildew, Fusarium wilt, soft rot, and Turnip mosaic virus [2]. Cultural, physical, biological, or chemical controls, or a combination of these controls, integrated pest management, are used for disease control [2, 3]. However, soil-borne phytopathogens such as Fusarium wilt or clubroot are hard to control by physical and chemical methods, and they can survive in the soil for many years in dormant conditions and become devastating when they find suitable host [2–4]. Thus, breeding the disease resistant cultivars of Brassica vegetables, especially against soil-borne phytopathogens, is the best way for effective disease control. Recently, some disease resistance genes (*R* genes) have been isolated in Brassica vegetables, and DNA marker assisted selection is applicable in some diseases [2, 3].

In this chapter, we focus on the Fusarium wilt and clubroot and present the breeding for these disease resistances in Brassica vegetables using DNA marker selection.

2. DNA marker selection for breeding Fusarium wilt disease resistant cultivars in Brassica vegetables

Fusarium species are highly host specific and comprise more than 120 *formae speciales* (f. sp.) further sub grouped into races [5, 6]. *Fusarium oxysporum* is considered as one of the top ten most devastating plant pathogens throughout the world and can infect approximately 150 of independent host plants or over, including economically important agricultural crops such as cabbage, tomato, onion, pepper, cucumbers, bananas, melons, cotton, etc. [5, 7–12]. Two *formae speciales* of *F. oxysporum* (f. sp. *conglutinans* and f. sp. *rapae*) mainly invade in Brassica vegetables [3, 13].

Fusarium wilt was first identified in the United States by Smith in the 1890s, and in the following decades it was subsequently found in Japan and several other countries [14, 15]. In recent years, Fusarium wilt has been overspread in China [16]. *F. oxysporum* f. sp. *conglutinans* infects Brassica vegetable roots (young roots are more vulnerable), and thereafter, it colonizes and blocks the xylem vessels by their growth leading to blockage of the water transport inside the plant. Finally, it leads to show the disease symptoms such as dull green to yellow green color of the leaves initially, yellowing, wilting, necrosis of leaf, defoliation, stunting, and death of seedling [17].

2.1 Traditional management

A number of traditional techniques have been adopted to manage Fusarium wilt disease. Crop rotation is effective to control Fusarium wilt disease [18], and soil solarization [19] and soil steam sterilization [20, 21] can suppress significantly the *F. oxysporum* f. sp. *conglutinans* population. Application of chemical fungicides such as prochloraz, carbendazim, and Bavistin is also used [22], but it is not strongly recommended to control Fusarium wilt [23, 24]. Chemical fumigants such as sodium azide, chloropicrin, and methyl bromide etc. are environmentally hazardous and most of them are not available nowadays. A few chemical fumigants may be available in commercial market in some countries, but it needs to be applied according to the sustainable regulations [25–27]. An alternative environment friendly method is biological control, but there are not any registered biological control agents for the Fusarium wilt in Brassica vegetables [27]. Combining different independent strategies are used for the more efficient control of Fusarium wilt. For example, combining the organic soil amendment (*Brassica carinata* defatted seed meals and compost) with a short period of soil solarization can significantly reduce both *F. oxysporum* f. sp. *conglutinans* and *F. oxysporum* f. sp. *raphani* [28].

2.2 Isolation of resistance genes

Most *R* genes encode proteins with leucine-rich repeats (LRR), a central nucleotide binding site (NBS) domain, and in the N-terminus a domain that contains homology to cytosolic domains of the Drosophila Toll or animal interleukin-1 receptors (TIR) (termed TIR–NBS–LRR) or a potential coiled coil (CC) domain (termed CC–NBS–LRR) [29–31]. *R* gene of *F. oxysporum* f. sp. *conglutinans* has been identified in *B. rapa* using transcriptome analysis focusing on differentially expressed putative *R* genes that have NBS, LRR, TIR, or CC motifs between Fusarium wilt resistant and susceptible lines [32]. Two TIR-NBS-LRR genes (Bra012688 and Bra012689), which located next to each other in the same transcriptional direction, have been identified as candidates of *R* gene (*FocBr1*), and presence and absence of these two genes were identical to the resistant and susceptible phenotypes, respectively, by inoculation test using F_2 population derived from crossing between susceptible and resistant lines [32, 33]. However, it has not been clarified which gene is *FocBr1* [32].

R gene of *F. oxysporum* f. sp. *conglutinans* (*FocBo1*) has also been identified in *B. oleracea* by genetic approach [34, 35], and candidate *R* gene in *B. oleracea* is ortholog of Bra012688 [35]. It suggests that Bra012688 could be *FocBr1*. The susceptible *B. oleracea* lines have mutations causing frame shift and there are several susceptible alleles [34–36], suggesting that mutations leading to susceptibility have occurred multiple times independently. In contrast, no mutations other than deletion leading to loss of function of *FocBr1* have been found in *B. rapa* [32, 33].

2.3 DNA marker selection system

Selection by inoculation test is labor-intensive and highly influenced by the environmental factors, and selection of suitable plants highly depends on the experience of breeders. In contrast, DNA marker selection is rarely affected by the environmental conditions. DNA marker selection also has merits that it can be performed at early developmental stages, can handle many samples, and can test multiple traits in a sample [37]. Identification of *R* gene or locus linked to *R* gene enables us to develop DNA marker for disease resistance [2, 3, 38].

As the susceptible allele (*focbr1–1*) of Fusarium wilt in *B. rapa* is due to deletion of *FocBr1*, a dominant DNA marker (Bra012688m), which confirms amplification of *FocBr1*, has been developed. This dominant DNA marker cannot distinguish the homozygous (*FocBr1*/*FocBr1*) and heterozygous (*FocBr1*/*focbr1–1*) alleles (**Table 1**). The SSR marker (SSR687int), which locates close to *FocBr1*, was identified, and we have confirmed this DNA marker can identify the heterozygous alleles in some lines (**Figure 1**). However, as there were several lines showing not identical to genotype information with resistance phenotypes by inoculation test, the genotypes determined by this DNA marker (SSR687int) and disease resistance by inoculation test must be confirmed before applying the DNA marker selection. To shorten the time required for PCR and to allow simultaneous determination of two dominant DNA markers of Bra012688 and Bra012689 (multiplex), the new DNA marker sets (YR688s and YR689s) were developed (**Table 1**).

In *B. oleracea*, three different susceptible alleles (*focbo1–1*, *focbo1–2*, and *focbo1–3*) were found [36]. As DNA marker sets covering these three susceptible alleles have been developed [36], it is necessary to select a DNA marker suitable for lines and will be available for breeder to use these primer sets (**Table 2**).

		Primer sequence	PCR condition	R	S	Ref.
Bra012688 (*FocBr1*)						
Bra012688m	F	AGTCGCTTGGAAGTCTGAGG	1 cycle of 94 °C for 3 min, 35 cycles of 94 °C for 30s, 58 °C for 30s, and 72 °C for 1 min, and final extension at 72 °C for 3 min.	A	NA	[32, 33]
	R	GAGCTAACCAACTATACATTGAACC				
YR688s	F	CTCCATCTGAGGATGGAAGTTGTACAAGCTCGGA	1 cycle of 94 °C for 2 min, 35 cycles of 94 °C for 30s and 68 °C for 30s, and final extension at 68 °C for 2 min. This primer set is used with YR689s for multiplex PCR.	A	NA	
	R	GCTCCGAATTCGAATTGGTGAATATCGCATACGAG				
SSR687int	F	CGTCAAAACCCTTTTGCCTA	1 cycle of 94 °C for 2 min, 35 cycles of 94 °C for 30s, 58 °C for 30s, and 72 °C for 30s, and final extension at 72 °C for 2 min. This marker is co-dominant DNA marker	LB	SB	
	R	CAAACGCTCGGTCCTGAAAT				
Bra012689						
Bra012689m	F	GCATCAAGGCAAAAATGTCA	1 cycle of 94 °C for 3 min, 35 cycles of 94 °C for 30s, 58 °C for 30s, and 72 °C for 1 min, and final extension at 72 °C for 3 min.	A	NA	[32, 33]
	R	CATTATAGTAGAACCCAAGTTGATCC				
YR689s	F	CCACTCAGATGTCGTTGAGAAGTCTGATACCATCG	1 cycle of 94 °C for 2 min, 35 cycles of 94 °C for 30s and 68 °C for 30s, and final extension at 68 °C for 2 min. This primer set is used with YR688s for multiplex PCR	A	NA	
	R	AGGAGACGACGACTGATCCACAAAGTGTGTATC				

R, resistance; S, susceptible; A, amplification of PCR product; NA, no amplification of PCR product; LB, large sized band; SB, small sized band.

Table 1.
DNA marker for predicting fusarium wilt resistance in Brassica rapa.

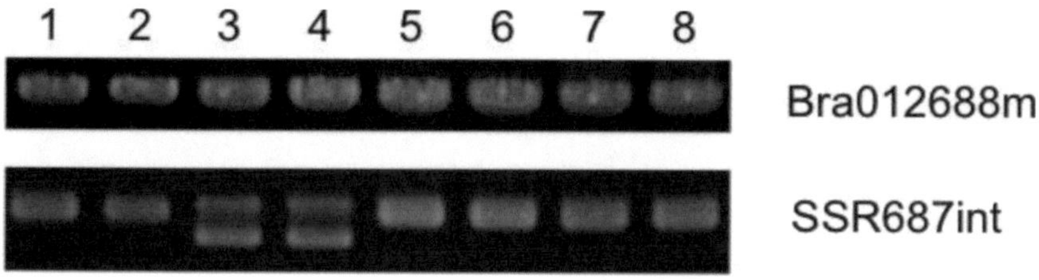

Figure 1.
Determining of Fusarium wilt resistance by DNA marker. A dominant DNA marker (Bra012688m) and co-dominant DNA marker (SSR687int) are used. Lane 3 and 4 show the heterozygosity of FocBr1 *and* focbr1–1 *alleles.*

3. DNA marker selection for breeding clubroot disease resistant cultivars in Brassica vegetables

Clubroot is also one of the most devastating diseases in Brassica vegetables and spreads almost all over the world [39]. Clubroot disease is caused by an obligate plant pathogen *Plasmodiophora brassicae,* which has distinct pathotypes or physiological races over the world [40, 41]. Plants infected by *P. brassicae* form clubs on roots, which interfere with the host plant's water and nutrients uptake. This interference leads to leaf yellowing, wilting, stunted growth, and death of the host plants.

3.1 Traditional management

Clubroot is quite difficult to control completely by the traditional methods due to the long survival spores of the *P. brassicae* in soil, their pattern of life cycle, and their pathotype specific infection, so that *P. brassicae* ultimately causes a broad diversity of virulence [3]. However, some traditional management system can control clubroot disease in some extent. Crop rotation with non-cruciferous plants can reduce the infestation of *P. brassicae* [18, 42], but cannot eliminate the *P. brassicae* completely [43, 44]. It is recommended not to grow any cruciferous plants on the infested site at least five to seven years.

Some biocontrol agents against *P. brassicae* such as *Bacillus subtilis*, *Streptomyces griseorube*, etc. are able to reduce the severity of clubroot infection [45, 46]. Soil sterilants like chloropicrin, diazomet, methyl, or ethylene dibromide etc. are effective to control clubroot [47]. Application of fungicides fluazinam and cyazofamid can effectively reduce the viability of resting *P. brassicae* spores and prevent infection [47]. However, the real fact is that these chemicals are not commercially approved for clubroot management since a long ago [47]. Integrated application of cultural or physical, chemical, and biocontrol agents can also be practiced for the more efficient management of *P. brassicae* [47, 48].

3.2 Isolation of resistance genes

In *B. rapa*, clubroot resistance is controlled by major dominant genes and pathotypes specific [3]. About 20 clubroot resistance loci have been identified, and highest numbers of clubroot resistance genes were found in chromosome A03 [3]. Two clubroot resistance genes (*CRa/b* and *Crr1a*) have been cloned and both genes encode a TIR-NB-LRR class R protein [49–51]. Recently, new clubroot resistance loci were found in a locus close to *Crr1a* in chromosome A08. A clubroot resistance locus, covering *CRs* gene, has been identified, and Bra020876 and Bra020918 have been identified as candidates of the *R* gene. Another clubroot resistance locus, covering

		Primer sequence	PCR condition	R	S	Ref.
Fusa-6	F	TGATGCAAGTGTGGTGACAA	1 cycle of 94 °C for 3 min, 35 cycles of 94 °C for 30s, 58 °C for 30s, and 72 °C for 1 min, and final extension at 72 °C for 3 min After PCR, *Hind* III digestion *FocBo1* vs. *focbo1–1*	D	ND	[61]
	R	CAATCGCTTCTTGCTTCTCC				
Fusa-4	F	ATCATGGGATCGAGAGAAGCCGCCC	1 cycle of 94 °C for 3 min, 35 cycles of 94 °C for 30s, 58 °C for 30s, and 72 °C for 1 min, and final extension at 72 °C for 3 min After PCR, *Eco* RI digestion *FocBo1* vs. *focbo1–1*	ND	D	[61]
	R	TAGCTTCATGCCATAGTCGTCCTGG				
#1	F	AGATTGTGCAATTAAACGCGACG	1 cycle of 94 °C for 3 min, 35 cycles of 94 °C for 30s, 58 °C for 30s, and 72 °C for 1 min, and final extension at 72 °C for 3 min After PCR, *Eco* RI digestion *FocBo1* vs. *focbo1–1*	ND	D	[36]
	R	CATCCTCAGATTCCAAGCACAAC				
#2	F	GAAGTTGGGTAAAGAAATTGTTCGTGC	1 cycle of 94 °C for 3 min, 35 cycles of 94 °C for 30s, 58 °C for 30s, and 72 °C for 1 min, and final extension at 72 °C for 3 min *FocBo1* vs. *focbo1–2*	SB	LB	[36]
	R	ATCCCAAGTTGATATCAGTAGGAAGAG				
#3	F	AATGGTTGCTCAATGAGAAGTATGC	1 cycle of 94 °C for 3 min, 35 cycles of 94 °C for 30s, 58 °C for 30s, and 72 °C for 1 min, and final extension at 72 °C for 3 min After PCR, *Eco* RI d-caps marker *FocBo1* vs. *focbo1–3*	ND	D	[36]
	R	GCCTCTGAAAGATCTGGAAAAAGAA				

R, resistance; S, susceptible; D, PCR products are digested; ND, PCR products are not digested; LB, large sized band; SB, small sized band.

Table 2.
DNA marker for predicting fusarium wilt resistance in Brassica oleracea.

Rcr9 gene, was also identified in chromosome A08, and Bra020936 has been identified as a candidate gene [52]. Two clubroot resistance loci, *Rcr*3 and *Rcr9*^{wa}, have been mapped in chromosome A08 being 1.17 Mb apart each other. Three genes, Bra020951, Bra020974, and Bra020979, have been identified as candidates of the *Rcr3* gene, and three genes related to immune-system-process (Bra020827, Bra020828, Bra020814) have been identified as candidates of the *Rcr9*^{wa} gene [53]. A clubroot resistance locus, covering *PbBrA08*^{Banglim} gene, was also detected on chromosome A08, where is near to *Crr1*, *CRs*, and *Rcr9* [54]. These reports suggest that they are not allelic, thus chromosome A08 covering these genes has an *R* gene cluster.

In *B. oleracea*, clubroot resistance is quantitative, and QTLs have been identified. Effect of each QTL is weak, and little progress in isolation of *R* gene or fine mapping of *R* gene in *B. oleracea* [3]. It clarifies resistant mechanism of clubroot in *B. oleracea* is polygenic nature where multiple clubroot loci combinedly responsible to the clubroot resistance [55].

3.3 DNA marker selection system

A breeding for clubroot resistance is much more complex compared with Fusarium wilt resistance due to the complexity of plant–pathogen interactions. A number of clubroot resistance locus has been identified by the different research groups in *B. rapa*, and this variation is due to the pathotype specific pathogenicity of *P. brassicae* [2, 3]. As *CRa*/*CRb* and *Crr1a* have been isolated, DNA markers in these genes have been developed. A dominant resistance (CRaim-T) and susceptible (Craim-Q) DNA marker set of *CRa* has been developed [49], and co-dominant indel marker (mCrr1a) of *Crr1a* has been developed (**Table 3**) [33, 50]. In *B. rapa*, as the other clubroot resistance genes have not been isolated, linkage DNA markers are developed (**Table 3**). Some clubroot resistant cultivars in Chinese cabbage have been produced by introducing a single gene for clubroot resistance from European turnip. However, there is a problem that the loss of resistance by the presences of multiple pathotypes of *P. brassicae* or arising new pathotypes has been found [56, 57]. Thus, the accumulation of multiple genes of clubroot resistance could make small risks to the breakdown of resistance [58, 59]. The introduction of DNA marker selection is essential for the simultaneous selection of multiple clubroot resistance genes. Furthermore, high-throughput genotyping system such as multiplex PCR could be useful. We have developed the multiplex DNA marker selection system (**Figure 2**). Indeed, the accumulation of three major clubroot resistance genes (*CRa*/*CRb*, *CRk*, and *CRc*) by DNA marker selection in Chinese cabbage represented the highly resistance against six isolates of *P. brassicae* [60]. A high clubroot resistant Chinese cabbage cultivar, 'Akimeki', was also developed by the accumulation of *Crr1a*, *Crr2*, and *CRa*/*CRb* genes by DNA marker selection [3].

In *B. oleracea*, it has also been compared by the independent and cumulative incorporation of the clubroot resistance locus to combat various isolates, where one major clubroot resistance gene (*PbBo(Anju)1*) accumulated independently, as well as combined with four minor clubroot resistance genes (*PbBo(Anju)2*, *PbBo(Anju)3*, *Pb-Bo(Anju)4*, *Pb-Bo(GC)1*). Accumulation of five clubroot resistance genes (one major and four minor clubroot resistance genes) represented the highest resistance against six *P. brassicae* isolates. Here, the major QTL, PbBo(Anju)1, is a main player for the resistance mechanism against *P. brassicae* and the introgression of other four minor clubroot resistance QTLs boosted up the resistance [55]. As this major clubroot resistance gene acts as repressive, heterozygous of this gene shows susceptibility to *P. brassicae*. Previously, we have tested using linkage DNA marker of *PbBo(Anju)1* in 35 cabbage F_1 cultivars in Japan, and 12 cultivars (34%) have homozygous of *PbBo(Anju)1* allele [61], suggesting that about 60%

		Primer sequence	PCR condition	R	S	Ref.
CRa/CRb						
CRaim-T	F	TATATTAATGATAAAGCAGAAGAAGAAA	1 cycle of 94 °C for 3 min, 35 cycles of 94 °C for 30s, 58 °C for 30s, and 72 °C for 1 min, and final extension at 72 °C for 3 min.	A	NA	[33, 49]
	R	AATGCGACTGAGAAAGTTGTAG				
Craim-Q	F	TGAAGAATGCGGGCTACGTCCTCTGAAATC		NA	A	
	R	GAAGTAGATGAACGTGTTTATTTTAGAAA				
CRbzhang						
TCR108	F	CGGATATTCGATCTGTGTTCA	1 cycle of 94 °C for 3 min, 35 cycles of 94 °C for 30s, 58 °C for 30s, and 72 °C for 1 min, and final extension at 72 °C for 3 min.	A	NA	[33, 62]
	R	AAAATGTATGTGTTTATGTGTTTCTGG				
Crr1a						
mCrr1a	F	CGATGACATGTCTGCCTTCT	1 cycle of 94 °C for 3 min, 35 cycles of 94 °C for 30s, 58 °C for 30s, and 72 °C for 1 min, and final extension at 72 °C for 3 min.	SB	LB	[33]
	R	TCTGAGATTCAACCGCTTCA				
CRc						
B50-C9	F	GATTCAATGCATTTCTCTCGAT	1 cycle of 94 °C for 3 min, 35 cycles of 94 °C for 1 min, 55 °C for 1.5 min, and 72 °C for 2 min, and final extension at 72 °C for 7 min.	A	NA	[60]
B50	R	CGTATTATATCTCTTTCTCCATCCC				
B50-6R	F	AATGCATTTTCGCTCAACC		NA	A	
B50	R	CGTATTATATCTCTTTCTCCATCCC				
CRk						
HC688–4	F	TCTCTGTATTGCGTTGACTG	1 cycle of 94 °C for 3 min, 35 cycles of 94 °C for 1 min, 60 °C for 1.5 min, and 72 °C for 2 min, and final extension at 72 °C for 7 min.	A	NA	[60]
HC688–6	R	ATATGTTGAAGCCTATGTCT				
HC688–4	F	TCTCTGTATTGCGTTGACTG		NA	A	
HC688–7	R	AAATATATGTGAAGTCTTATGATC				
Crr2						

		Primer sequence	PCR condition	R	S	Ref.
BRMS-096	F	AGTCGAGATCTCGTTCGTGTCTCCC	1 cycle of 94 °C for 1 min, 35 cycles of 94 °C for 1 min, 40 °C for 1 min, and 72 °C for 1 min, and final extension at 72 °C for 4 min.	LB	SB	[63]
	R	TGAAGAAGGATTGAAGCTGTTGTTG				
Crr3						
OPC11–2S	F	GTAACTTGGTACAGAACAGCATAG	1 cycle of 94 °C for 30s, 45 cycles of 94 °C for 30s, 40 °C for 1 min, and 72 °C for 2 min, and final extension at 72 °C for 7 min.	LB	SB	[64, 65]
	R	ACTTGTCTAATGAATGATGATGG				

R, resistance; S, susceptible; A, amplification of PCR product; NA, no amplification of PCR product; LB, large sized band; SB, small sized band.

Table 3.
DNA marker for predicting clubroot resistance in Brassica rapa.

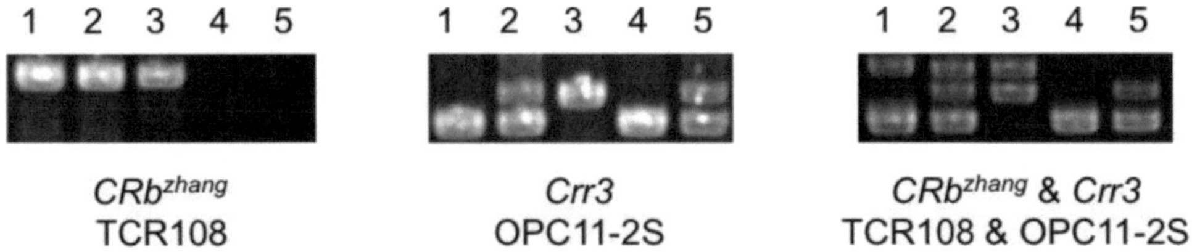

Figure 2.
Determining of clubroot resistance by DNA marker. A dominant DNA marker (TCR108) or co-dominant DNA marker (OPC11–2S) is used in left or middle panel, respectively. In right panel, two primer sets (TCR108& OPC11–2S) were used for the simultaneous selection of multiple clubroot resistance genes.

of cabbage cultivars could be susceptible to *P. brassicae*. One reason for clubroot resistance breeding in *B. oleracea* being behind in *B. rapa* is due to the theoretically impossible to introduce recessive resistance gene by backcrossing with inoculation test. However, DNA marker selection can overcome this problem as co-dominant DNA marker can distinguish the heterozygosity and homozygosity of *PbBo(Anju)1* allele, suggesting that introduction of DNA marker selection is indispensable in *B. oleracea*. However, the current DNA markers are linkage markers, making it difficult to use them universally. Thus, it will be necessary to develop gene markers based on mutations that cause susceptibility, and this will require the isolation of clubroot resistance genes in *B. oleracea*.

4. Perspective

Both Fusarium wilt and clubroot are the serious disease for Brassica vegetables. Breeders are trying to develop the resistant lines for the both diseases by DNA marker assisted breeding. It has already been successfully developed Fusarium wilt and clubroot resistant lines. However, a Fusarium wilt resistant line can be infested by the clubroot or vice versa, while the clubroot has the virulence complexity. It is quite difficult to inoculate the multiple pathogens/races in an individual plant, while resistant breeding independently for each disease will make a further issue. DNA marker-based selection will enables us to overcome the mentioned issue. It has already found an association between a Fusarium wilt resistance allele and clubroot susceptible allele in *B. napus*, but their recombination was also reported [66]. It is necessary to identify the possible linkage between the genes responsible for the Fusarium wilt and clubroot diseases in Brassica vegetables. A Fusarium wilt resistance gene (*FocBr1*) is located on the region covering *CR* genes (*CRa/CRb*, *Rcr1*, *Crr3*, and *CRk*) with a physical distance approximately 2 Mb in chromosome A03 [3]. Recombination of two genes has been found [33], thus we can accumulate Fusarium and clubroot resistant alleles. In *B. oleracea*, *FocBo1* is close to a minor clubroot QTL in chromosome C06, but they are not closely linked each other [35, 55, 61, 67]. A linkage between dissimilar resistance loci can allow to inherit the resistance genes both for Fusarium wilt and clubroot, which can lead us for the development of resistant cultivars for both diseases.

Acknowledgements

This chapter includes results supported by grants from Project of the NARO Bio-oriented Technology Research Advancement Institution (Research program on development of innovation technology).

Conflict of interest

The authors declare no conflict of interest.

Author details

Mst Arjina Akter[1,2], Hasan Mehraj[1*], Takeru Itabashi[3], Tomoe Shindo[3], Masaaki Osaka[3], Ayasha Akter[1,4], Naomi Miyaji[1], Naoki Chiba[3], Junji Miyazaki[5] and Ryo Fujimoto[1*]

1 Graduate School of Agricultural Science, Kobe University, Kobe, Japan

2 Department of Plant Pathology, Faculty of Agriculture, Bangladesh Agricultural University, Mymensingh, Bangladesh

3 Miyagi Prefectural Agriculture and Horticulture Research Center, Natori, Japan

4 Department of Horticulture, Faculty of Agriculture, Bangladesh Agricultural University, Mymensingh, Bangladesh

5 Agriculture Victoria Research Division, Department of Jobs, Precincts and Regions, AgriBioscience, Bundoora, Victoria, Australia

*Address all correspondence to: hmehraj34@stu.kobe-u.ac.jp and leo@people.kobe-u.ac.jp

References

[1] Lv H, Miyaji N, Osabe K, Akter A, Mehraj H, Shea DJ, Fujimoto R. The importance of genetic and epigenetic research in the *Brassica* vegetables in the face of climate change. In: Kole C, editor. Genomic Designing of Climate-Smart Vegetable Crops. Cham: Springer; 2020. p. 161-255. DOI: 10.1007/978-3-319-97415-6_3

[2] Lv H, Fang Z, Yang L, Zhang Y, Wang Y. An update on the arsenal: Mining resistance genes for disease management of Brassica crops in the genomic era. Horticulture Research. 2020;7:34. DOI: 10.1038/s41438-020-0257-9

[3] Mehraj H, Akter A, Miyaji N, Miyazaki J, Shea DJ, Fujimoto R, Doullah MA. Genetics of clubroot and Fusarium wilt disease resistance in Brassica vegetables: The application of marker assisted breeding for disease resistance. Plants. 2020;9:726. DOI: 10.3390/plants9060726

[4] Shlevin E, Mahrer Y, Katan J. Effect of moisture on thermal inactivation of soilborne pathogens under structural solarization. Phytopathology. 2004;94:132-137. DOI: 10.1094/PHYTO.2004.94.2.132

[5] Armstrong GM, Armstrong JK. Formae speciales and races of *Fusarium oxysporum* causing wilt diseases. In: Nelson PE, Toussoun TA, editors. Fusarium: Diseases, Biology and Taxonomy. University Park, Pennsylvania: Pennsylvania State University Press; 1981. p. 391-399.

[6] Lievens B, Houterman PM, Rep M. Effector gene screening allows unambiguous identification of *Fusarium oxysporum* f. sp. *lycopersici* races and discrimination from other *formae speciales*. FEMS Microbiology Letters. 2009;300:201-215. DOI: 10.1111/j.1574-6968.2009.01783.x

[7] MacHardy WE, Beckman CH. Vascular wilt Fusaria: Infection and pathogenesis. In: Nelson PE, Toussoun TA, editors. Fusarium: Disease, Biology, and Taxonomy. University Park, Pennsylvania: Pennsylvania State University Press; 1981. p. 365-390.

[8] Fourie G, Steenkamp ET, Ploetz RC, Gordon TR, Viljoen A. Current status of the taxonomic position of *Fusarium oxysporum formae specialis cubense* within the *Fusarium oxysporum* complex. Infection, Genetics and Evolution. 2011;11:533-542. DOI: 10.1016/j.meegid.2011.01.012

[9] Dean R, Van Kan JAL, Pretorius ZA, Hammond-Kosack KE, Di Pietro A, Spanu PD, Rudd JJ, Dickman M, Kahmann R, Ellis J, Foster GD. The top 10 fungal pathogens in molecular plant pathology. Molecular Plant Pathology. 2012;13:414-430. DOI: 10.1111/j.1364-3703.2011.00783.x

[10] Khan N, Maymon M, Hirsch AM. Combating *Fusarium* infection using *Bacillus*-based antimicrobials. Microorganisms. 2017;5:75. DOI: 10.3390/microorganisms5040075

[11] Sun Y, Wang M, Li Y, Gu Z, Ling N, Shen Q, Guo S. Wilted cucumber plants infected by *Fusarium oxysporum* f. sp. *cucumerinum* do not suffer from water shortage. Annals of Botany. 2017;120:427-436. DOI: 10.1093/aob/mcx065

[12] de Lamo FJ, Takken FLW. Biocontrol by *Fusarium oxysporum* using endophyte-mediated resistance. Frontiers in Plant Science. 2020;11:37. DOI: 10.3389/fpls.2020.00037

[13] Edel-Hermann V, Lecomte C. Current status of *Fusarium oxysporum formae speciales* and races.

Phytopathology. 2019;109:512-530. DOI: 10.1094/PHYTO-08-18-0320-RVW

[14] Smith EF. The fungus infection of agricultural soils in the United States. Scientific American Supplementary. 1899;48:19981-19982. DOI: 10.1038/scientificamerican11181899-19981supp

[15] Gillman JC. Cabbage yellows and the relation of temperature to its occurrence. Annals of the Missouri Botanical Garden. 1916;3:25-84. DOI: 10.2307/2990072

[16] Xing M, Lv H, Ma J, Xu D, Li H, Yang L, Kang J, Wang X, Fang Z. Transcriptome profiling of resistance to *Fusarium oxysporum* f. sp. *conglutinans* in cabbage (*Brassica oleracea*) roots. PLoS ONE. 2016;11:e0148048. DOI: 10.1371/journal.pone.0148048

[17] Gaetán SA. Occurrence of Fusarium wilt on canola caused by *Fusarium oxysporum* f. sp. *conglutinans* in Argentina. Plant Disease. 2005;89:432-432. DOI: 10.1094/PD-89-0432C

[18] Shafique HA, Sultana V. Ehteshamul-Haque S, Athar M. Management of soil-borne diseases of organic vegetables. Journal of Plant Protection Research. 2016;56:221-230. DOI: 10.1515/jppr-2016-0043

[19] Ramirez-Villapudua J, Munnecke DE. Control of cabbage yellow (*Fusarium oxysporum* f. sp. *conglutinans*) by solar heating of field soils amended with dry cabbage residues. Plant Disease. 1987;71:217-221. DOI: 10.1094/PD-71-0217

[20] Afek U, Orenstein J. Disinfecting potato tubers using steam treatments. Canadian Journal of Plant Pathology. 2002;24:36-39. DOI: 10.1080/07060660109506968

[21] Panth M, Hassler SC, Baysal-Gurel F. Methods for management of soilborne diseases in crop production. Agriculture. 2020;10:16. DOI: 10.3390/agriculture10010016

[22] Song W, Zhou L, Yang C, Cao X, Zhang L, Liu X. Tomato Fusarium wilt and its chemical control strategies in a hydroponic system. Crop Protection. 2004;23:243-247. DOI: 10.1016/j.cropro.2003.08.007

[23] Uesugi Y. Fungicide Classes: Chemistry, Uses and Mode of Action. In: Hutson D, Miyamoto J, editors. Fungicidal activity: Chemical and biological approaches to plant protection. New York: John Wiley & Sons; 1998. p. 23-56.

[24] Ma YT, Fan HF, Gao YQ, Li H, Zhang AL, Gao JM. Natural products as sources of new fungicides (I): synthesis and antifungal activity of acetophenone derivatives against phytopathogenic fungi. Chemical Biology & Drug Design. 2013;81:545-552. DOI: 10.1111/cbdd.12064

[25] Colla P, Garibaldi A, Gullino ML. A review and critical analysis of the European situation of soilborne disease management in the vegetable sector. Phytoparasitica. 2012;40:515-523. DOI: 10.1007/s12600-012-0252-2

[26] Colla P, Garibaldi A, Gullino ML. Consequences of European pesticide policies enforcement in soil disinfestation sector. Acta Horticulturae. 2014;1044:363-366. DOI: 10.17660/ActaHortic.2014.1044.48

[27] Arie T. *Fusarium* diseases of cultivated plants, control, diagnosis, and molecular and genetic studies. Journal of Pesticide Science. 2019;44:275-281. DOI: 10.1584/jpestics.J19-03

[28] Gilardi G, Demarchi S, Moizio M, Gullino ML, Garibaldi A. Disease management strategies against Fusarium wilt of basil and rocket by combining amendments and soil

solarization. Acta Horticulturae. 2014;1044:319-324. DOI: 10.17660/ActaHortic.2014.1044.41

[29] Meyers BC, Kozik A, Griego A, Kuang H, Michelmore RW. Genome-wide analysis of NBS–LRR-encoding genes in Arabidopsis. The Plant Cell. 2003;15:809-834. DOI: 10.1105/tpc.009308

[30] Joshi RK, Nayak S. Functional characterization and signal transduction ability of nucleotide-binding site-leucine-rich repeat resistance genes in plants. Genetics and Molecular Research. 2011;10:2637-2652. DOI: 10.4238/2011.October.25.10

[31] Marone D, Russo MA, Laidò G, De Leonardis AM, Mastrangelo AM. Plant nucleotide binding site-leucine-rich repeat (NBS-LRR) genes: active guardians in host defense responses. International Journal of Molecular Sciences. 2013;14:7302-7326. DOI: 10.3390/ijms14047302

[32] Shimizu M, Fujimoto R, Ying H, Pu ZJ, Ebe Y, Kawanabe T, Saeki N, Taylor JM, Kaji M, Dennis ES, Okazaki K. Identification of candidate genes for fusarium yellows resistance in Chinese cabbage by differential expression analysis. Plant Molecular Biology. 2014;85:247-257. DOI: 10.1007/s11103-014-0182-0

[33] Kawamura K, Kawanabe T, Shimizu M, Nagano AJ, Saeki N, Okazaki K, Kaji M, Dennis ES, Osabe K, Fujimoto R. Genetic characterization of inbred lines of Chinese cabbage by DNA markers; towards the application of DNA markers to breeding of F_1 hybrid cultivars. Data in Brief. 2016;6:229-237. DOI: 10.1016/j.dib.2015.11.058

[34] Lv H, Fang Z, Yang L, Zhang Y, Wang Q, Liu Y, Zhuang M, Yang Y, Xie B, Liu B, Liu J, Kang J, Wang X. Mapping and analysis of a novel candidate Fusarium wilt resistance gene *FOC1* in *Brassica oleracea*. BMC Genomics. 2014;15:1094. DOI: 10.1186/1471-2164-15-1094

[35] Shimizu M, Pu ZJ, Kawanabe T, Kitashiba H, Matsumoto S, Ebe Y, Sano M, Funaki E, Fujimoto R, Okazaki K. Map-based cloning of a candidate gene conferring Fusarium yellows resistance in *Brassica oleracea*. Theoretical and Applied Genetics. 2015;128:119-130. DOI: 10.1007/s00122-014-2416-6

[36] Sato M, Shimizu M, Shea DJ, Hoque M, Kawanabe T, Miyaji N, Fujimoto R, Fukai E, Okazaki K. Allele specific DNA marker for fusarium resistance gene *FocBo1* in *Brassica oleracea*. Breeding Science. 2019;69:308-315. DOI: 10.1270/jsbbs.18156

[37] Collard BYC, Mackill DJ. Marker-assisted selection: an approach for precision plant breeding in the twenty-first century. Philosophical Transactions of the Royal Society B. 2008;363:557-572. DOI: 10.1098/rstb.2007.2170

[38] Hirani AH, Gao F, Liu J, Fu G, Wu C, Yuan Y, Hou J, Duncan R, Li G. Transferring clubroot resistance from Chinese cabbage (*Brassica rapa*) to canola (*B. napus*). Canadian Journal of Plant Pathology. 2016;38:82-90. DOI: 10.1080/07060661.2016.1141799

[39] Hirani AH, Li G. Understanding the genetics of clubroot resistance for effectively controlling this disease in Brassica species. In: Hany El-Shemy, editor. Plants for the Future. IntechOpen; 2015. DOI: 10.5772/60936.

[40] Kageyama K, Asano T. Life cycle of *Plasmodiophora brassicae*. Journal of Plant Growth Regulation. 2009;28:203-211. DOI: 10.1007/s00344-009-9101-z

[41] Tommerup IC, Ingram DS. The life-cycle of *Plasmodiophora brassicae* Woron. In *Brassica* tissue cultures and in intact roots. New Phytologist.

1971;70:327-332. DOI: 10.1111/j.1469-8137.1971.tb02531.x

[42] Wallenhammar AC, Johnsson L, Gerhardson B. Agronomic performance of partly clubroot-resistant spring oilseed turnip rape lines. Journal of Phytopathology. 2000;148:495-499. DOI: 10.1046/j.1439-0434.2000.00536.x

[43] Faggian R, Strelkov SE. Detection and Measurement of *Plasmodiophora brassicae*. Journal of Plant Growth Regulation. 2009;28:282-288. DOI: 10.1007/s00344-009-9092-9

[44] Daniel C, Hommes M, Koller M. Plant protection in organic production of *Brassica* vegetables and oilseed rape. IOBC-WPRS Bulletin. 2016;118:110-115.

[45] Peng G, McGregor L, Lahlali R, Gossen BD, Hwang SF, Adhikari KK, Strelkov SE, McDonald MR. Potential biological control of clubroot on canola and crucifer vegetable crops. Plant Pathology. 2011;60:566-574. DOI: 10.1111/j.1365-3059.2010.02400.x

[46] Wang J, Huang Y, Lin S, Liu F, Song Q, Peng Y, Zhao L. A strain of *Streptomyces griseoruber* isolated from rhizospheric soil of Chinese cabbage as antagonist to *Plasmodiophora brassicae*. Annals of Microbiology. 2012;62:247-253. DOI: 10.1007/s13213-011-0253-2

[47] Donald C, Porter I. Integrated control of clubroot. Journal of Plant Growth Regulation. 2009;28:289-303. DOI: 10.1007/s00344-009-9094-7

[48] Donald EC, Porter IJ, Faggian R, Lancaster RA. An integrated approach to the control of clubroot in vegetable Brassica crops. Acta Horticulturae. 2006;706:283-300. DOI: 10.17660/ActaHortic.2006.706.33

[49] Ueno H, Matsumoto E, Aruga D, Kitagawa S, Matumura H, Hayashida N. Molecular characterization of the *CRa* gene conferring clubroot resistance in *Brassica rapa*. Plant Molecular Biology. 2012;80:621-629. DOI: 10.1007/s11103-012-9971-5

[50] Hatakeyama K, Suwabe K, Tomita RN, Kato T, Nunome T, Fukuoka H, Matsumoto S. Identification and characterization of *Crr1a*, a gene for resistance to clubroot disease (*Plasmodiophora brassicae* Woronin) in *Brassica rapa* L. PLoS ONE. 2013;8:e54745. DOI: 10.1371/journal.pone.0054745

[51] Hatakeyama K, Niwa T, Kato T, Ohara T, Kakizaki T, Matsumoto S. The tandem repeated organization of NB-LRR genes in the clubroot-resistant *CRb* locus in *Brassica rapa* L. Molecular Genetics and Genomics. 2017;292:397-405. DOI: 10.1007/s00438-016-1281-1

[52] Laila R, Park JI, Robin AHK, Natarajan S, Vijayakumar H, Shirasawa K, Isobe S, Kim HT, Nou IS. Mapping of a novel clubroot resistance QTL using ddRAD-seq in Chinese cabbage (*Brassica rapa* L.). BMC Plant Biology. 2019;19:13. DOI: 10.1186/s12870-018-1615-8

[53] Karim MM, Dakouri A, Zhang Y, Chen Q, Peng G, Strelkov SE, Gossen BD, Yu F. Two clubroot-resistance genes, *Rcr3* and *Rcr9*^{wa}, mapped in *Brassica rapa* using bulk segregant RNA sequencing. International Journal of Molecular Sciences. 2020;21:5033. DOI: 10.3390/ijms21145033

[54] Choi SR, Oh SH, Chhapekar SS, Dhandapani V, Lee CY, Rameneni JJ, Ma Y, Choi GJ, Lee SS, Lim YP. Quantitative trait locus mapping of clubroot resistance and *Plasmodiophora brassicae* pathotype Banglim-specific marker development in *Brassica rapa*. International Journal of Molecular Sciences. 2020;21:4157. DOI: 10.3390/ijms21114157

[55] Tomita H, Shimizu M, Doullah MA, Fujimoto R, Okazaki K. Accumulation of quantitative trait loci conferring

broad-spectrum clubroot resistance in *Brassica oleracea*. Molecular Breeding. 2013;32:889-900. DOI: 10.1007/s11032-013-9918-9

[56] Kuginuki Y, Yoshikawa H, Hirai M. Variation in virulence of *Plasmodiophora brassicae* in Japan tested with clubroot-resistant cultivars of Chinese cabbage (*Brassica rapa* L. ssp. *pekinensis*). European Journal of Plant Pathology. 1999;105:327-332. DOI: 10.1023/A:1008705413127

[57] Hatakeyama K, Fujimura M, Ishida M, Suzuki T. New classification method for *Plasmodiophora brassicae* field isolates in Japan based on resistance of F_1 cultivars of Chinese cabbage (*Brassica rapa* L.) to clubroot. Breeding Science. 2004;54:197-201. DOI: 10.1270/jsbbs.54.197

[58] Neik TX, Barbetti MJ, Batley J. Current status and challenges in identifying disease resistance genes in *Brassica napus*. Frontiers in Plant Science. 2017;8:1788. DOI: 10.3389/fpls.2017.01788

[59] Strelkov SE, Hwang SF, Manolii VP, Cao T, Fredua-Agyeman R, Harding MW, Peng G, Gossen BD, Mcdonald MR, Feindel D. Virulence and pathotype classification of *Plasmodiophora brassicae* populations collected from clubroot resistant canola (*Brassica napus*) in Canada. Canadian Journal of Plant Pathology. 2018;40:284-298. DOI: 10.1080/07060661.2018.1459851

[60] Matsumoto E, Ueno H, Aruga D, Sakamoto K, Hayashida N. Accumulation of three clubroot resistance genes through marker-assisted selection in Chinese cabbage (*Brassica rapa spp. pekinensis*). Journal of the Japanese Society for Horticultural Science. 2012;81:184-190. DOI: 10.2503/jjshs1.81.184

[61] Kawamura K, Shimizu M, Kawanabe T, Pu Z, Kodama T, Kaji M, Osabe K, Fujimoto R, Okazaki K. Assessment of DNA markers for seed contamination testing and selection of disease resistance in cabbage. Euphytica. 2017;213:28. DOI: 10.1007/s10681-016-1821-0

[62] Zhang T, Zhao Z, Zhang C, Pang W, Choi SR, Lim YP, Piao Z. Fine genetic and physical mapping of the *CRb* gene conferring resistance to clubroot disease in *Brassica rapa*. Molecular Breeding. 2014;34:1173-1183. DOI: 10.1007/s11032-014-0108-1

[63] Suwabe K, Tsukazaki H, Iketani H, Hatakeyama K, Fujimura M, Nunome T, Fukuoka H, Matsumoto S, Hirai M. Identification of two loci for resistance to clubroot (*Plasmodiophora brassicae* Woronin) in *Brassica rapa* L. Theoretical and Applied Genetics. 2003;107:997-1002. DOI: 10.1007/s00122-003-1309-x

[64] Saito M, Kubo N, Matsumoto S, Suwabe K, Tsukada M, Hirai M. Fine mapping of the clubroot resistance gene, *Crr3*, in *Brassica rapa*. Theoretical and Applied Genetics. 2006;114:81-91. DOI: 10.1007/s00122-006-0412-1

[65] Hirai M, Harada T, Kubo N, Tsukada M, Suwabe K, Matsumoto S. A novel locus for clubroot resistance in *Brassica rapa* and its linkage markers. Theoretical and Applied Genetics. 2004;108:639-643. DOI 10.1007/s00122-003-1475-x

[66] Rahman H, Franke C. Association of fusarium wilt susceptibility with clubroot resistance derived from winter *Brassica napus* L. 'Mendel'. Canadian Journal of Plant Pathology. 2019;41: 60-64. DOI: 10.1080/07060661.2018.1505780

[67] Pu ZJ, Shimizu M, Zhang YJ, Nagaoka T, Hayashi T, Hori H, Matsumoto S, Fujimoto R, Okazaki K. Genetic mapping of a fusarium wilt resistance gene in *Brassica oleracea*. Molecular Breeding. 2012;30:809-818. DOI: 10.1007/s11032-011-9665-8

Chapter 9

Brassica-Aphid Interaction: Modulated Challenges and Sustainable Approach for Management

S.A. Dwivedi, Lelika Nameirakpam and Ajay Tomer

Abstract

Insect pests act as main barrier in enhancing yield potential of Brassica crops. *Lipaphis erysimi* is considered as one of the most destructive insect species in mustard production due to its voracious type feeding and multiplication. Therefore application of insecticide is inevitable for cultivation of cruciferous crops, although systemic insecticides has been found to be suitable for management of aphid, despite of high cost, residual effect and ecological ramification have necessitated the application of bio and botanical insecticides as novel approach and are recorded significant in research. Aphids having exclusively viviparous parthenogenesis type reproduction from January to March month with the completion of eight generations are helpful in quick mass multiplication. Natural enemies *Coccinella* spp., Syrphid larvae and bio-pesticide found effective in suppress aphid numbers. Manipulation in sowing dates of mustard crop provides good yield and less incidence of aphid which is proved through research. Lack of environmental resistant varieties has dispensed toward non feasibility of conventional breeding approaches for developing aphid-resistant Brassica. Although application of genetic engineering plan has resulted in moderate success in development of aphid resistance, so far commercialization of such genetically modified crops has not conceivable, intimate the necessity of further insights in to host plant and aphid communication to form effective approach against aphid resistance. Therefore in this chapter the components involved in Brassica aphid communication are highlighted and present statuses and problem in aphid management are discussed.

Keywords: aphid, ecological factors, entomopathogenic fungus, predators, resistance varieties, systemic insecticide, yield loss

1. Introduction

Rape seeds-mustard act as a major valuable oilseed and create key commencement of utilisation of oil and cake for feeding purpose of human as well as animal respectively. It has crucial status in Indian recession. India ranked 2nd in the production of mustard among all oil seed crops followed by China [1]. Mustard shared total 26% of production of oil seed in India. Main component of mustard is oil (32–40%) and protein (15–17%) Oilseeds as dietary food on priority basis and stored as raw material in agro industry are used to prepare various commodity such as cosmetics, detergents, laxatives, soaps, lubricants, apart from it have excellent medical and therapeutic significant. Application of recent package of practices with the cultivation of high yielding varieties enhances production of mustard. Rape seed mustard are highly susceptible to incidence of several pests like mustard aphid (*L.erysimi* Kalt), painted bug (*Bagrada picta*), sawfly (*Athalia proxima*),) leaf minor *(Phytomyza atricornis*) and flea beetle (*Phyllotreta cruciferae)*, among, *L. erysimi* is most destructive deliberate pest of mustard. Aphid act as key crop pest due to its damaging capability of target crop in recent cropping pattern, It acts as alarming arthropod and spreaded globally including temperate and subtropical territory. Aphids suck phloem and chlorophyll tissues from tender portion of plants and causing qualitative and quantitative yield-limiting factor. Infestation of aphid decreases in the yield by reducing no. of pods/plant, no. of grains/pod and oil content within grains (**Figure 1**). Aphid has overcome the barrier of glucosinolates becoming involved in self protection against insects those feed on the phloem content and sequestering these compounds arresting them within body. Abiotic components such as temperature, light, moisture, wind velocity etc. express clear response on incidence as well as multiplication of aphid population, among them, temperature played significant role in multiplication of aphid and air current and rain fall were noted as significant factors for survival as well as dispersion of aphid [2]. Occurrence and intensity of aphid mainly gets in trouble by climatic factors. This pest remains active throughout the growth period of crop up to pod drying by consuming liquid content from tender vegetative portion, floweral parts and siliqua of mustard. Immature and adults stage feed on succulent vegetative and pod formation stages of crop resulting in stunted growth, wither floral parts and grains undeveloped in siliqua. Infested leaves become wrapped and discoloured, brownish marking develops on vegetative portion and show wilting symptom. *L.erysimi*

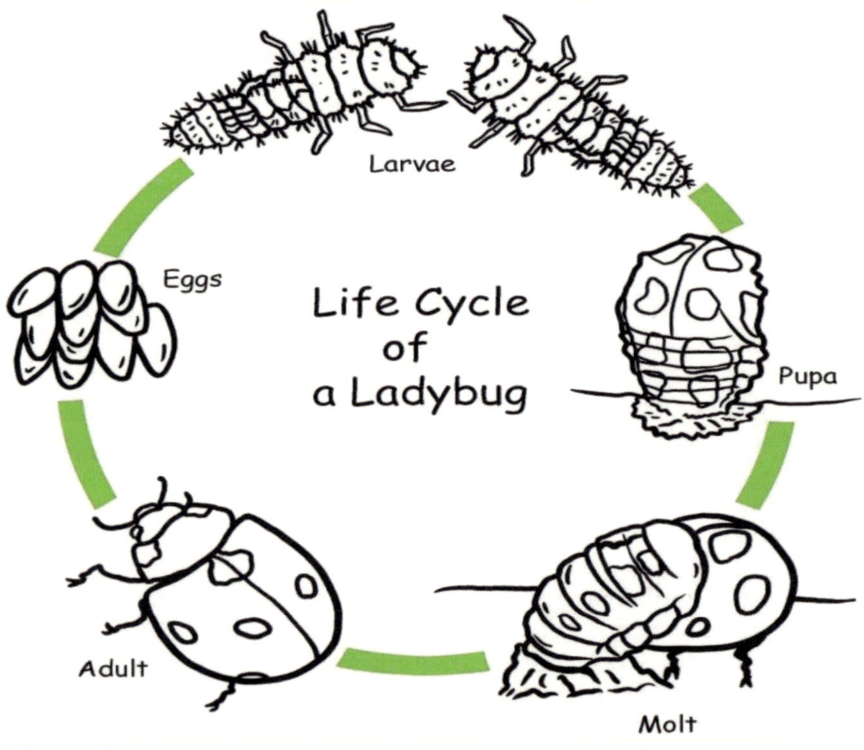

Figure 1.
Life cycle of predator Lady bird beetle Coccinella septempunctata.

release sticky sweet substance which develops sooty moulds as a result vegetative portion appears black patches and faces photosynthesize inhibition [1]. Mustard aphid caused 9%–95% production losses. In India at different locality aphid caused tremendous 83% loss in rapeseed and mustard 91.3% and 34.68% at Kanpur, 59.49% at Pant Nagar, 72.61% at Ludhiana, 29.43% at Navgaon. Regarding management aphid farmers rely up on the application of synthetic chemical that creates harmful condition like residual content of toxic substance, forming resistance against target pests and indiscriminate use of such chemical causes environmental pollutions, mortality of bioagents etc. To avoid such adverse things, finding out aphid resistance or tolerant cultivars is the best effective practices for management of target pest. Mustard aphid can be managed by release of natural enemy. Among them effective bioagents are like, syrphid flies, *Syrphus confrater* (Weid.), *Syrphus balteatus* (Deg.), *Ischiodon scutellaris* (Fab.). *Coccinella septempunctata* is most effective insect feeder on various types of plant lice that recorded as successful bio agent of *L. erysimi*. Sprinkler irrigation helpful to wash aphid colony those attached to the apical shoot of plant and reduce aphid population by mixing them in soil. Irrigation for 2–3 times is found effective in aphid management and is economically sound. Several sustainable approaches are discussed in this article with the help of researchers' results regarding management of aphid in mustard crop.

2. Host plant resistance as effective phenomenon for controlling aphid

Crops infested by aphids are those having good sap content [3]. Consecutive selection of a plant, aphid required to adjust with it to obtain benefit from target crops. Pest consumes liquid content as its feeding material from phloem of plant via inserting stylets [4]. Plants external arrangement as well as manufactured complex substances of plants perform key role for safety of plant against aphid. External structure like, waxy content on leaf, hardness of fingernail skin, availability of spines and trichome affect aphid for selection of target portion of plant [5]. Further, leaves having alternative metabolites, healthful condition of fluid content of plant portion act as target host by plant lice [6]. Phytophagous crucifixion as well as essentiality of plants are altered with changeable climatic condition that at last ramification for their communications. [7], Increase temperature, carbon dioxide, moisture stress, environmental pollutant generally SO_2, NO as well as NO_2 enormously alter population of aphid to select its suitable target host [8]. Correspondingly, be concerned with development of aphid and their collaboration with other biotic additionally decided link with aphid and target host plant [9].

3. Nourishing mechanism of aphid on target host

Aphid changes their size by moulting process in nymphal form body that depend up on the nourishment gain from target host. Inside complexity of all harmful arthropods of mustard, aphid has the ability to bear carotenoid shade from normally in selected hosts [10]. Plant lice species does not impel toward other plant canopy as their host plant. On their selected target they attacked on generally all tender parts of plant, like vegetative, floweral part, branches and pod. Plant cell sap is suck by modified piercing and sucking type mouthparts of aphids, mouth parts of aphid are modified as needle like structure stylets combination that slices target tissue of plant to insert in phloem site and concurrently stylet penetrate in to the phloem. Aphid form two particular types of spit, protein as well as jellifying thick saliva around the stylet helpful to create an intercellular course in phloem for the purpose of penetrating stylets [11], next sorts of saliva discharge occurred to takeoff filter through stylet into the vascular structure of target host. Aphid release sugar rich material recognisable as honeydew that enhances the improvement of dirty form in the monetary patches of plants and curtail the nature of item [12]. Yet, honeydew sweet in nature attract ants for spare them from normal foes of aphid. Continuation ways of aphid about 20–40 days; its higher increase rate acts as its life assurance for maintain their population in crop ecosystem by providing protection from natural enemies.

4. Reproduction pattern of aphid

Aphid shows both sexual and asexual type of reproduction capability along with comparatively simple reproductive adjustment. On the basis of availability of host plant aphid expresses either autoecious (No change in host, monoecious) or heteroecious type of life history. Mainly aphid completes monoecious life cycle, by spending entire life on single host plant [13] but on the other hand, only 10% aphid is noted as heteroecious by completing their single life cycle on different hosts [14]. On the basis of environmental situation, aphid is capable to produce of nymphs or eggs at different time of year, it may be holocyclic means completing life cycle changing between parthenogenesis or sexual reproduction or anholocyclic means incomplete life cycle expressing only parthenogenesis but no sexual reproduction pattern life cycle followed by aphid [15]. In favourable condition aphids promote both type of life cycle. In holocyclic life cycle at low temperature eggs on primary host hatched in spring, developed in to winged mother (fundatrices), which quickly convert parthenogenesis or viviparous type of reproduction promoting wingless female population shortly. With increase of temperature wingless female gave birth of new apterous generation of aphid. In cold condition apterous aphid promoted into alate form, a few of which were males participated in sexual reproduction by mating with female and returned on primary for oviposition [16]. At the beginning of spring season hatching of these eggs occurred for recycling of life (**Figure 2**). Males are completely absent only asexual reproduction is recorded in anholocyclic life cycle. Viviparous females gives birth only female aphid parthenogenetically throughout the year (**Figure 2**). Mustard aphids are located mostly in various geographical locations, where overwintering oviposition process almost completely absent, it shows parthenogenetic type reproduction by entire year [17].

The adult females deposited eggs on tender leaves and shoot and go through an advancement of hatching. Such growth and development of plant lice with no preparation produce their little girl aphid. This structure develops via parthenogenesis type reproduction in hilly area [18]. It has affection for selection of host plant for deposition of egg mass in hilly area. Host attributes like, genetically modification, external appearance, physiological structure, engineering, appropriation, thickness of vegetative portion and physical signs are considered by plant feeder as well as aphids for proper selection of their ovipositional place [19]. In the mid year time frame they pick woody hosts for optional or agricultural crops, including vegetable harvests of families Chenopodiaceae, Compositae. Cucurbitaceae, Cruciferae and Solanaceae [20]. Yet, in the ephemeral crops aphid deposited their eggs mass on floral parts or young branches near to floweral parts [21]. Natural as well as synthetic characters of flowers of target crops alter oviposition of aphid. Female adults find out safety as well as mechanical assist in the deposition of eggs due to them select elongated floweral parts generally. Main parts of leaf having alleco-synthetic admixture as well as lipids can beside create oviposition [22]. Crop volatilise beside supporting in the reproductive improvement help in the and release of sex pheromones

Figure 2.
Infestation of Lipaphis erysimi *on mustard crop.*

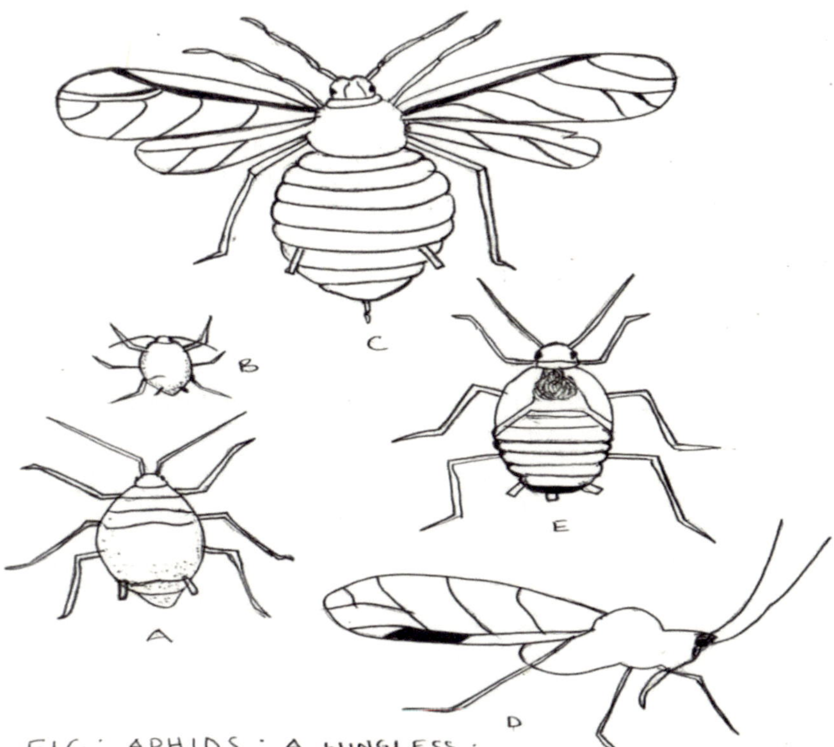

Figure 3.
Stages of mustard aphid Lipaphis erysimi.

by female aphid [23]. In plain region, *L. erysimi* reproduces entirely by viviparous parthenogenesis type reproduction from January to March month, in this particular period, the aphid completed, on an average eight generations (**Figure 3**) [24, 25].

5. Effects of temperature and drought condition on growth and multiplication of aphids

Temperature play an important role in managing wing spread, divergence, improvement as well as evolution of life stages in aphid [26]. In summer season aromatic plants provide best quality food comparison with wooded plant. Plant lice can overthrow the command forced at high temperature from dislocate themselves from that territory's host plant to other target host [27]. Increase the strength of aphid colony in crop ecosystem depends upon the optimum range of temperature. In different experiment, it was clear that occurrence and intensity of aphid were directly related on temperature as well as warm moist cloudy weather on mustard [28]. There are several acceptances that water compression approach in the recurrence of some phytophagous arthropods [29]. Aphid depends on the with balanced water pressure on plants [30]. Thus, aphid tries to move another place from their disturbed place and starts feeding on host crops where development of population easily takes placed with reduction of yield.

5.1 Factors influencing the selection and modification of target crops by aphid

Plant lice are one of the valuable agricultural destructive arthropods in crop production related with 4500 species globally. Its short life cycle completed within month, with high fecundity facilitates them to continue their destruction on crops by mass multiplication and maintaince population in the field. It acts as vector of transmitting viral diseases. Application of chemical to manage target pest population within field crops has harmful issues in as creating environmental pollution and health hazard. Regular use of synthetic molecules creates resistance in target pest as well as changes status of small population of pest in to major problem. Eco-friendly pest management practices can provide useful way for reduction of aphid population from field crop. Proper handling of crop ecosystem segment supplies excellent choice to avoid harmful effect of pesticide application. Reciprocal action of plant lice with their host plant is a basic principles for protect environment from chemical pollutant. Target crop of pests that provide shelter as well as nutritive food, aphids are phytophagous in nature dependent on various agricultural crops to complete life cycle [3]. After finding suitable host plant, aphid accommodate with it to take required nutrient from plant. They ingest liquid content as food material from phloem region of host by inserting their stylet [4]. External arrangement as well as synthetic molecule on crops is the first part of defence of plant to counter the attack of aphid such as waxy coating on upper part of leaf, hard integument, availability of ridges and trichome alter plant lice to search target crops [5]. Nutritional status and water availability within cell sap and secondary metabolites interfere in searching suitable target crops by aphid [6]. Phytophagous pest activity as well as attributes of host is affected by the modification of climatic condition that ultimately disturbs their interactions. [7], Exalted temperature, CO_2, moisture stress as well as ecosystem pollutants like SO_2, NO and NO_2 show significant impact on aphid multiplication and finding their target host crops [8]. In further, nature of damage as well as birth rate of plant lice and its intercommunication with another living organism are helpful to decide the relation among them [9]. Simple correlation with meteorological parameters revealed that among the abiotic factors (Temperature, relative humidity and rainfall), temperature had the biggest impact in enhancement as well as maintenance of aphid populace. The appearance of *Coccinella* spp. and the larvae of *Syrphid flies* are positively correlated with temperature, while there was negative correlation with the occurrence of mustard aphid *Lipaphis erysimi*. There is positive correlation between the population of aphid and relative humidity [24].

5.2 Comparable study on life table of *L.erysimi* on alternate host

Canola acts as important cash crop in Iran. *L. erysimi* is key pests of cruciferous crops globally having 10–90% damaging capability relaying on the harshness of attack on target host [31, 32]. Aphid is capable to damage on leaf, flower and fruits of canola [33]. Regarding management of aphid application of chemical pesticides causes a lot of adverse effects including toxic effects on natural enemies, outbreak of secondary pest, contamination of food web and residues creating problem on the aspect of health hazard of living organism in ecosystem [1]. To find out substitute chemical in pest management, use of bioagents is an effective tool [34]. Work on Life stages makes it easy to consider the population dynamics of insects and provide information about reproduction, survivality and development [35–38]. Lot of research work studies have appraised the effect of various *Brassica* germplasm on demographic limitation of *Plutella xylostella* (L.) [36, 39–41], *Chromatomya horticola* Goureau [42], *Myzus persicae* [42], *Thrips tabaci* [43], *Brevicoryne brassicae* L. [44, 45]. Additionally, response of several canola germplasm on various life stages of *L. erysimi* were already studied [32, 46]. Including the multiplication factors of aphid and its natural enemies on canola host at several nitrogen fertiliser treatments [34].

5.3 Function of effector protein in spreading of aphid

It is considered that available protein in aphid saliva acts as effector proteins with specific disparate function that combine to stop immune process of the target crop formation of effective colony, new approach of bioinformatics and proteomics instrument applied for identification scant strength of effectors in aphid [47–49]. Few of them effectors express excepted work like as cell wall degradation with enzyme (Amylases, pectinases, glucanases) or detoxification (peroxidases, phenol oxidases, oxidoreductase) but generally this effector was recorded as dissimilarity to protein with known work [48].

5.4 Communication through signal response in host following aphid infestation

Endogenous signalling molecule of host crop performs a significant role in the management of protective response against attack of phytophagous. Communications between the plant hormones like as gibberellic acid (GA), jasmonic acid (JA), abscisic acid (ABA), salicylic acid (SA), hydrogen peroxide (H_20_2) and nitric oxide(NO) creates a complex interrelated structure where all component influence each other by both synergistic and inhibitory communication proceeded to a protective mechanism [4]. Aphid like as *Brevicoryne brassicae, Myzus persicae* has been reported to defeat host crop by introducing resistance via manipulating of cross communication in between signalling molecules through promoting of SA- dependent pathway as well as concurrently down promoting JA-dependent pathway [50].

6. Biogical aspect as well as sustainable potential of three effective bio control agents against *L. erysimi*

Management of aphid's natural enemies such as, Ladybird beetle, *Coccinella septumpunctata* (Linnaeus), Syrphid flies, *Episyrphus viridaureus* (Wiedemann),

Betasyrphus isaaci (Bhatia) perform significant role in mid altitude hills of Meghalaya. Basic speciality of natural enemies and functional status against target pest is very much essential to utilise them judiciously. Consequently, the biological aspect regarding consuming strength of *C. septempunctata* and syrphid flies were studied in lab condition, to get their effectiveness, strength as well as more benefits in reduction of aphid population [51–54]. *Lipaphis erysimi* (Kalt.) was found to be parasitized by ten hymenopterous parasites, belonging to two families, five genera. Out of these parasities *Diaeretus rapae* and *Aphidius spp*. play significant role in reducing aphid population. [55] *M. anisopliae* and *B. bassiana* were the most effective with less toxicity against Ladybird beetle and syrphid fly by continuously increasing population after application [56].

6.1 *Coccinella septempunctata*

Female adult deposited yellow coloured eggs in group near about 26–45. Hatching duration 3.5 ± 0.5 days to be recorded, growth and size of the larva enhanced with each successive ecdysis. Total grub duration was recorded 26 ± 3 days.. Grey to black in colour with external orange pupa was observed of *C. septempunctata*. The size of the adult and pupa approximated the same (**Figure 4**). The pupal duration was recorded 7.5 ± 1.5 days, longevity of female adult was 131.5 ± 1.5 days as well as fecundity was 357.45 ± 22.41 eggs [57, 58]. Adult beetle on an average consumed 95 aphids per day [59] adult consumed 339 aphids and larva 540 aphids (**Table 1**) [61].

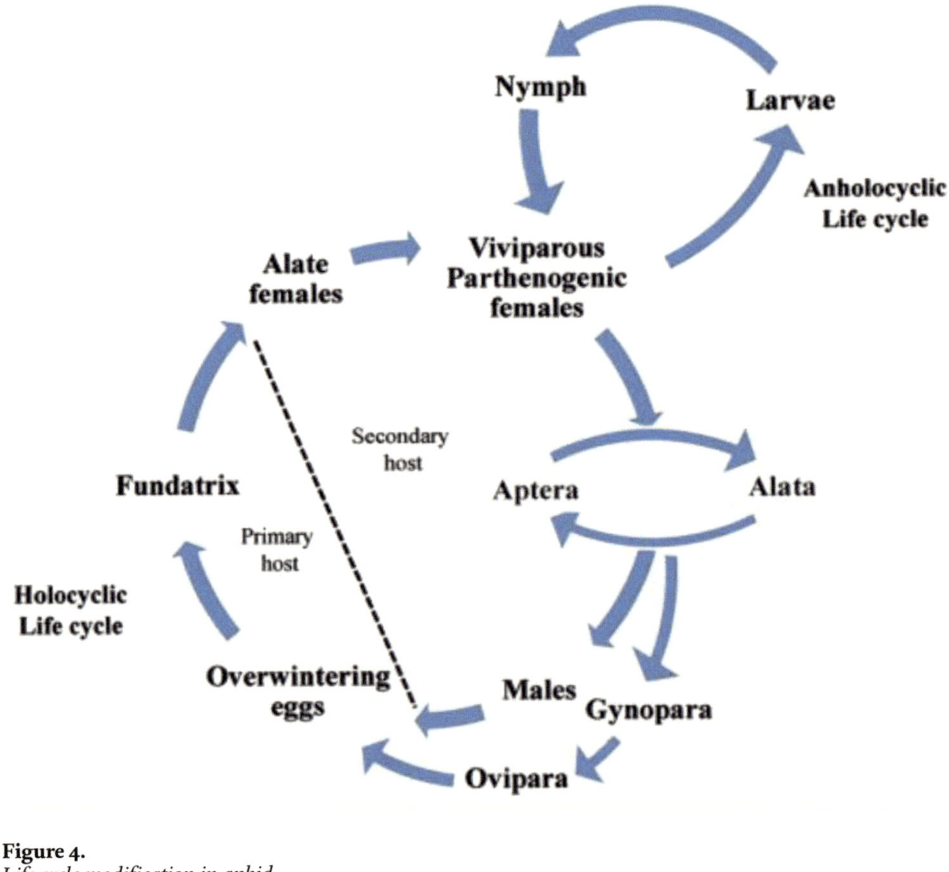

Figure 4.
Life cycle modification in aphid.

Parameter	Predators		
	Coccinella septempunctata	***Episyrphus viridaureus***	***Betasyrphus isaaci***
Incubation period	3.5 ± 0.5 days	03 ± 0.5 days	3 ± 1 days
Larval period	26 ± 3 days	22 ± 1.5 days	21 ± 1.5 days
First instar	3.5 ± 0.5 days	12.9 ± 1.0 days	13 ± 0.5 days
Second instar	7.5 ± 1.5 days	4.1 ± 0.5 days	3.90 ± 1.0 days
Third instar	6.5 ± 0.5 days	5.0 ± 1.0 days	4.0 ± 0.5 days
Fourth instar	8.5 ± 1.0 days	—	—
Pupal period	7.5 ± 1.5 days	7 ± 1 days	8 ± 1 days
Adult longevity	31.5 ± 1.5 days	14 ± 1.5 days	13 ± 1 days
Life cycle	68.5 ± 6.5 days	47 ± 2 days	41 ± 2 days
Fecundity	357.45 ± 22.41. No/female	45.0 ± 16.8. No/female	31.2 ± 13.6. No/female

Source: [60].

Table 1.
Biological attributes of three predators of mustard aphids under laboratory conditions.

6.2 *Episyrphus viridaureus*

Near or within colony of aphid single eggs deposition occurred by *E. Viridaureus.* White colour and oblong in shape eggs hatching was recorded up to 3 ± 0.5 days. Immature stage completed three larval instars. Intrusting, apodus larvae of *E. viridaureus* had a permeable body, internal organs clearly visible. Life span of larvae was recorded to be 22 ± 1.5 days. Creamy as well as pear frame, tapered at the one side of pupae had 7 ± 1 day duration. Longevity of adult female was a 14 ± 1.5 day with fecundity was 45.0 ± 16.8 eggs. Total life history was completed in 47 to 49 days. f *E. balteatus* was recorded to take 21.2 days to completes its life cycle having larval duration of 7.6 days (**Table 1**) [62].

6.3 *Betasyrphus isaaci*

Greyish in colour as well as oblong shaped eggs deposited by adult female had incubation duration 3 ± 1 days. Larval period completed within 21 ± 1.5 days having three larval instars. 8 ± 1 days were recorded as pupal period. Longevity of adult female was 13 ± 1 days as well as laid 31.2 ± 13.6 eggs (**Table 1**).

6.4 Consumption capability of predators on aphid

The study on these predators, feeding capability on plant lice noticed that last grub instar devoured highest aphids than earlier instar grub and enhance each consecutive instars. Such capability of natural enemies' grub of *C. septumpunctata* was observed higher than both the species of syrphid flies. Individual adults of lady bird beetle feed on an average of 81.55 ± 15.34 aphids per daily and ultimately feed on 2691.00 ± 533 aphids during mature stage. Both grub and adult stages of it are predatory in nature and therefore it was recorded most superior predator of mustard aphid. One adult feed near about 4312 ± 537.74 aphids in a lifespan; which is much more than *E. viridaureus* (416.67 ± 6.76 aphids) and white fly (338 ± 7.89 aphids). Maximum feeding occurred during final instar of grub which could be

Stages	Daily consumption of aphids per day (Mean ± SE)			Consumption per life stage (Mean ± SE)		
	Coccinella septempunctata	*Episyrphus viridaureus*	*Betasyrphus isaaci*	*Coccinella septempunctata*	*Episyrphus viridaureus*	*Betasyrphus isaaci*
First instar	20.42 ± 00.42	07.30 ± 0.08	06.51 ± 0.17	081.67 ± 0.33	095.00 ± 1.51	084.67 ± 1.30
Second instar	35.00 ± 00.99	23.58 ± 0.22	23.75 ± 0.29	315.00 ± 1.34	094.00 ± 4.30	095.00 ± 4.80
Third instar	65.48 ± 01.27	45.53 ± 1.09	39.58 ± 1.46	458.33 ± 1.29	227.67 ± 0.95	158.33 ± 1.79
Fourth instar	85.11 ± 01.39	—	—	766.00 ± 1.78	—	—
Adult	81.55 ± 15.34	Free living	Free living	2691.00 ± 533	Free living	Free living
Total cons umption				4312 ± 537.74	416.67 ± 6.76	338.00 ± 7.89

Source: [60].

Table 2.
Feeding potential of three major predators of Lipaphis erysimi.

associated with modification of mouth structure as well as excellent metabolism than early instars. This result provided support to several outcomes on feeding capability of different syrphid and *coccinellids* [1, 63–65]. The first to fourth instar of grub of lady bird beetle feed on 21.43, 46.90, 72.61, and 102.60 aphids daily, respectively [1]. The feeding capability on prey of *Episyrphus spp*. enhanced slowly with the growth of grub [64]. Observation regarding the feeding potential of white fly is not available in the existing literature, however, reported that the first, second and third instar of another closely related syrphid, *B. serarius* feeds on 11.5, 44.75 and 232.5 aphids daily (**Table 2**) [65].

7. Occurrence and management of mustard aphid through cultural practices

Thirty-eight insect pest incidences are recorded on mustard crop in India. In the country among them aphid acts as key pest in mustard growing region. Nymphs and adults both stages of aphid damaged crop by sucking liquid food material from the leaves, flowers as well as siliquae making the qualitative and quantitive loss in yield. Aphid reduced 35.4 to 96% yield loss, 30.9% weight loss and 2.75 per cent oil loss in mustard [66–69].

7.1 Date of sowing

The occurrences of *L. erysimi* as well as its population build up were recorded at full flowering stage and full pod setting stage of the crops. The yield of various varieties was recorded at harvest. Rapeseed-mustard varieties sown during first and third week of October, minimum level of aphid infestation, while those sown in first and third week of November, were infested heavily, Among the varieties, the gobhi sarson (HPN-1) was highly susceptible to the aphid attack, while *B. carinata* (HPC-1) was least infested as compared to other varieties. Varieties sown early provided greater yield, while Varuna and HPC-1 gave the higher yield than the rest, irrespective of sowing date [70]. The *L. erysimi* population was minimum in crops sown on 10thOctober and maximum in crops sown on 24thNovember where average aphid population was 40.70 aphids/10 cm twigs. Indian mustard sown on 10thOctober successfully evaded the infestation of the 2 insect pests during the study [71]. Significantly least aphid population of 7.3 and 7.4 aphids/10 cm apical shoot on the seasonal total emergence to maturity was recorded on early sowing. Variety Rohini (15th October) provided the effective combination having less aphid population but higher yield, 58.6 and 60.4 aphids/10 cm apical shoot and seed yield, 1670.7 and 1915.1 kg/ha [2].

7.2 Utilisation of aphid resistant variety

Application of resistance cultivar acts as eco-friendly way to control aphid infestation on Brassica crops. For development of resistant variety utilisation of conventional breeding techniques required lot of time and repetition due to deficiency of resistant component in cultivated as well as wild relative of Brassica. In recent screening of two wild type Brassica varieties (*B. fruticulosa* and *B. montana*) followed by breeding chance of *B. juncea* showing heritable introgession against resistance of aphid in lab condition [72]. Based on pooled mean of aphid infestation index (0–5 rating scale), genotypes were classified to different grade of resistance. Out of 65 genotypes, six genotypes viz., NDR-05-1, RW-2-2, ONK-1, NRCKR-299, Kiran and T-27 were categorised as highly resistant, 16 genotypes were found as resistant, 21 genotypes were found moderately resistant, 13 genotypes were graded as susceptible and remaining nine genotypes were highly susceptible. Three *Brassica* genotypes (NRCKR-299, Kiran and T-27) were found consistently as highly resistant at both full flower and pod stages [73, 74]. On the basis of aphid infestation index at the time of flowering as well as siliqua development, it was observed that varieties Varuna and Vaibhav were susceptible to aphid infestation. Uravasi, Maya, Vardan, Ashirvad and Pitambari were noted as fairly resistant to aphid while Rohini showed resistance to aphid incidence [75]. Avoidable mustard production loss due to L. *erysimi* were checked in four cultivar of Karan rai, Ethiopian mustard as comparative with Indian mustard Varuna [76].

7.3 Balanced application of fertilisers

Combined utilisation of biofertilizers, growth retardant and compost can therefore be employed for regulating crop metabolism and physiological responses resulting in enhanced crop growth and protection against pathogens and pest [77].

7.4 Role of yellow sticky trap in aphid management

Performance of yellow sticky trap and imidacloprid 17.8% SL was assessed on farmer's field through front line demonstrations. The per cent increase in the yield under demonstration technology was 18.52% and 26.99% over the farmer's practices [78]. Monitoring of alate aphid initial average population ranged from 0.93 to 19.42 aphids per trap and attained to peak at interval relay upon the climatic factors from 9th to 12th standard week [79]. The initial average population ranged from 0.2 to 0.6 aphids per trap and came to peak alternately relaying upon the climatic factors during 7th to 10th standard week [80].

8. Application of entomopathogenic fungus in management of aphid

Lot of commercial fungal biopesticides with several brand names as well as formulations are available as agro-product globally [81]. The perverted entomopathogenic fungi *Beauveria bassiana* (Balsamo) *Vuillemin* and *Metarhizium anisopliae* (Metschnikoff) Sorokin are bioagents of a wide range of soft bodied insects including aphids, mealy bugs and arachnids; both fungi have a cosmopolitan distribution [82, 83]. *Lecanicillium (Verticillium) lecanii* (Zimm.) Zare & Gams has been used against greenhouse whitefly, thrips and aphids [84–86]. Similarly, *Paecilomyces fumosoroseus*, *P. farinosus and P. lilacinus* have been reported as entomopathogenic on a variety of insect pests [87, 88]. Very little information is available on the use of indigenous entomopathogenic fungi for the control of insect pests in Pakistan [84, 85]. A local strain *of M. anisopliae* was applied against cabbage aphid *Brevicoryne brassicae* L. This strain has also been screened for its compatibility with insecticides. Similarly, two local strains *of M. anisopliae* were used against *Coptotermes heimi* Wasmann [89]. The present report describes the efficacy of exotic and indigenous strains of *M. anisopliae, Paecilomyces lilacinus, Lecanicillium lecanii* and *B. bassiana* against the mustard aphid. Among entomopathogenic biopesticides *M. anisopliae* (83.23%) was found to be the most effective against mustard aphid followed by *B. bassiana* (78.33%) and *B. thuringiensis* (73%). Bio-pesticides can be used as a potential candidate for integrated pest management against mustard aphid after field efficacy [90]. Biological control of crop pests and diseases has been found to play significant role in reducing the over reliance on chemical pesticides.

9. Botanical pesticides

The crude aqueous extracts from *Ageratum conyzoides* (L.), *Parthenium hysterophorus* (L.), *Lantana camera* (L.), *Solanum nigrum* (L.), *Cannabis sativa* (L.), *Calotropis gigantean* (L.), *Livistona chinensis* (Jacq.), *Cassia angustifolia* (Mill.) were checked for its insecticidal as well as repellent activity against *M. persicae* (Sulzer) and *Brevicoryne brassicae* (Linnaeus). Repellent activity was inversely related to concentration of plant extract [91]. The antioxidant activities of different fraction of the methanolic extracts were indicated in the range of 69.08–84.89%. Thirty-four leaf extracts as well as *Azadirachta indica* were checked against healthy aphids kept in petri plates. It was observed that all the treatments show insecticidal properties versus aphid but the extract from *Chrysanthemum*, *Calotropis procera* noted result at par with *A. indica.* The other plant extracts *Zingiber offcinale, Ageratum conyzoides*, *Lantana camera*, Pinus roxburghii, *Allium sativum*, *Ricinus communis*, *Cymbopogon citrates* and *Hevea brasiliensis* yielded excellent outcomes [92] showing in **Table 3**.

S. No.	Local Name	Scientific Name	Parts used	Per cent morality of aphid
1	Adrak	*Zingiber officinale*	Leaves	22.20
2	Bael	*Aegel marmelos*	Leaves	14.43
3	Neela phulnu	*Ageratum conyzoides*	Leaves	29.96
4	Panch phuli	*Lantana camera*	Leaves	22.16
5	Banna	*Vitex negundo*	Leaves	13.30
6	Curry leaf	*Murraya koengii*	Leaves	6.66
7	Bougainvillea	*Bougainvillea glabra*	Leaves	9.86
8	Mint	*Mentha spicata*	Leaves	8.86
9	Bhang	*Cannabis sativa*	Leaves	22.20
10	Neem	*Azadirachta indica*	Leaves	35.43
11	Simal	*Bombax ceiba*	Leaves	15.50
12	Camphor	*Cinnamomum camphora*	Leaves	6.63
13	Morphanki	*Thuja orientalis*	Leaves	6.63
14	Datura	*Datura stramonium*	Leaves	4.40
15	Congress grass	*Parthenium hysterophorus*	Leaves	9.96
16	Pines	*Pinus roxburghii*	Leaves	26.63
17	Bamboos	*Bambusa arundinacea*	Leaves	4.40
18	Darek	*Melia azedarach*	Leaves	9.96
19	Jungle chulai	*Amaranthus spinosus*	Leaves	1.22
20	Amla	*Pylllanthus emblica*	Leaves	8.86
21	Harrar	*Terminalia chebula*	Leaves	18.86
22	Ak	*Calotropis procera*	Leaves	32.20
23	Gul-e—Daudi	*Chrysanthum coronarium*	Leaves	41.06
24	African Marigold	*Tagetus erecta*	Leaves	17.76
25	Burweed	*Xanthium strumarium*	Leaves	6.63
26	Kinnow	*Citrus sinensis*	Leaves	19.96
27	Garlic	*Allium sativum*	Leaves	25.53
28	Soybean	*Glycine max*	Leaves	17.73
29	Castor	*Ricinus communis*	Leaves	23.30
30	Talhi	*Delbergia sissoo*	Leaves	18.86
31	Lemon grass	*Cymbopogon citrates*	Leaves	26.63

S. No.	Local Name	Scientific Name	Parts used	Per cent morality of aphid
32	Jambolan	*Syzygium cumini*	Leaves	16.66
33	Rubber plant	*Hevea brasiliensis*	Leaves	22.20
	CD (P = 0.05)			5.8

Source: Srivastava & Guleria, (2003).

Table 3.
Evaluation of various plant-extracts against mustard aphid, Lipaphis erysimi.

10. Conclusion

In this chapter it can be concluded that aphid acts as dominant among all pest of mustard crop having 10–90% damaging capability with a significant reduction of yield. To avoid indiscriminate application of synthetic pesticides those show harmful effect on beneficial organism and application of eco-friendly management practices should be employed. However we will require extending of dynamics communication between host plant resistance as well as biological control with target pest in relation to changing climatic condition.

Acknowledgements

We would like to acknowledge the improvement of this manuscript made through comments by reviewer.

Author contributions

Listed authors have made substantial direct and intellectual contribution for this review article as well as approved it to submit for publication.

Author details

S.A. Dwivedi[1]*, Lelika Nameirakpam[1] and Ajay Tomer[2]

1 Department of Entomology, School of Agriculture, Lovely Professional University, Phagwara Punjab, India

2 Institute of Plant and Microbial Biology, Academia Sinica, Taiwan

*Address all correspondence to: dwivedi.sunilkumar46@gmail.com

References

[1] Singh, K. & Singh, N.N. (2013) Preying capacity of different established predators of the aphid *Lipaphis erysimi* (Kalt.) infesting rapeseed-mustard crop in laboratory conditions. Plant Protection Science,49(**2**): 84-88.

[2] Dwivedi, S.A. & Singh, R.S. (2019) The occurrence and intensity of mustard aphid *Lipaphis erysimi* (Kalt.) on dates of sowing in different varieties respect to yield parameter, Journal of Entomological Research 43 (4) : 491-498.

[3] Kumar, S. (2019) Aphid-Plant Interactions: Implications for Pest Management. In: Vegetation-Natural and Cultivated Vegetation in a Changing World. *Plant Communities and Their* Environment 1-19.

[4] Morkunas, I. & Gabryś B (2011) Phytohormonal signaling in plant responses to aphid feeding. Acta Physiological Plant, 33: 2057-2073.

[5] War, A.R., Paulraj, M.G, Ahmad, T, Buhroo, A.A. & Hussain, B. (2012) Mechanisms of plant defence against insect herbivores. Plant Signal Behaviour, 7: 130620.

[6] Powell, G., Tosh, C.R., & Hardie, J. (2006) Host plant selection by aphids: behavioral, evolutionary, and applied perspectives. Annual Review of Entomology, 51: 309-330.

[7] Awmack, C.S., Harrington, R., & Lindroth, R.L. (2004) Aphid individual performance may not predict population responses to elevated CO_2or O_3. Global Change Biology 10: 1414-1423.

[8] Sun, Y., Guo, H. & Ge F (2016) Plant–aphid interactions under elevated CO2: some cues from aphid feeding behavior. Front Plant Science, 7: 502.

[9] Watanabe, S., Murakami, Y. & Hasegawa, E. (2018) Effects of aphid parasitism on host plant fitness in an aphid-host relationship. PLoS One 13: 1-12

[10] Moran, N.A. & Jarvik, T. (2010) Lateral transfer of genes from fungi underlies carotenoid production in aphids. Science *328*(5978), 624-627

[11] Guerrieri, E. & Digilio M.C. (2008) Aphid-plant interactions: a review. Journal of Plant Interaction, 3: 223-232.

[12] Binazzi, A. & Scheurer, S. (2009) Atlas of the honeydew producing conifer aphids of Europe In: Aracne, Rome, Italy.

[13] Dixon, A.F.G. (1985) Structure of aphid populations. Annual review of entomology. 30(1): 155-174.

[14] Moran, N.A. (1992) The evolution of aphid life cycles. Annual Review of Entomology 37(1): 321-348

[15] Agarwala, B.K. (2007) Phenotypic plasticity in aphids (Homoptera: Insecta): components of variation and causative factors. Current Science-Bangalore. 93(3): 308.

[16] Bhatia, V., Uniyal, P.L. & Bhattacharya, R. (2011) Aphid resistance in *Brassica* crops: challenges, biotechnological progress and emerging possibilities. Biotechnology advances. 29(6): 879-888.

[17] Conrad, J. (2009) The aphid life cycle, the backyard nature. http://www.backyardnature.net/aphid_lc.html

[18] Dedryver, F. (2013) The genetics of obligate parthenogenesis in an aphid species and its consequences for the maintenance of alternative reproductive modes. Heredity (Edinb) 110: 39.

[19] War, A.R., Taggar, G.K., Hussain, B., Taggar, M.S. & Nair, R.M. (2018)

Plant defence against herbivory and insect adaptations. AoB Plants 10: 1-19.

[20] Powell, G., & Hardie, J. (2001) The chemical ecology of aphid host alternation: How do return migrants find the primary host plant? Applied Entomology of Zoology, 36: 259-267.

[21] Holopainen, J.K. (2002) Aphid response to elevated ozone and CO2, Entomologia Exp- erimentalis Applicata, 104: 137-142

[22] Stadler, B., Dixon, A.F.G., Kindlmann, P. (2002) Relative fitness of aphids: effects of plant quality and ants. Ecology Letters, 5: 216-222.

[23] Hurley, J., Takemoto, H., Takabayashi, J. & McNeil, J.N. (2014) Host Plant Volatiles and the Sexual Reproduction of the Potato Aphid, *Macrosiphum euphorbiae* Insects 5: 378-392.

[24] Dwivedi, S.A., Singh, R.S. & Gharde, S.K.(2018a) Populations build up of mustard aphid and their natural enemies in relation to biotic and abiotic factors, Plant Archives 18(2)2495-2500

[25] Dwivedi, S.A., Gharde, S.K. & Singh R.S. (2018b) Biology of Mustard Aphid, *Lipaphis erysimi* (Kalt.) in Laboratory Condition, Annals of Biology 34 (2): 167-169

[26] Guschina, I.A. & Harwood, J.L. (2006) Mechanisms of temperature adaptation in poikilotherms, Febs Letters, 580: 5477-5483.

[27] Loxdale, H.D. & Balog, A. (2018) Aphid specialism as an example of ecological-evolutionary divergence. Biological Reviews, 93: 642-57. 41.

[28] Rao, B.B., Rao, V.U.M., Nair, L., Prasad, Y.G. &Ramaraj, A.P. (2013) Assessing aphid infestation in Indian mustard (*Brassica juncea* L.) under present and future climate scenarios. *Bangla* Journal of Agricultural Research, 38: 373-387.

[29] Foote, N.E., Davis, T.S., Crowder, D.W., Bosque-Pérez, N.A., & Eigenbrode, S.D. (2017) Plant water stress affects interactions between an invasive and a naturalized aphid species on cereal crops. Environmental Entomology 46: 609-616.

[30] Liu, D., Dai, P., Li, S., Ahmed, S.S., & Shang, Z, (2018) Life-history responses of insects to water-deficit stress: a case study with the aphid *Sitobion avenae*. Biomedical Chromatography Ecology, 18: 17

[31] Liu, S. S., Wang, X. G., Wu, X. J., Shi, Z. H., Chen, Q. H. & Hu, H. X. (1997) Population fluctuation of aphids on crucifer vegetables in *Hangzhou suburbs*. (Chinese) *Acta Applied* Ecology, 8: 510-514.

[32] Rana, J. (2005) Performance of *Lipaphis erysimi* (Homoptera: Aphididae) on Different *Brassica Species* in a Tropical Environment. Journal of Pesticide Science, 78: 155-160.

[33] Goggin, F. L. (2007) Plant-aphid Interactions: Molecular and Ecological Perspectives. Current Opinion, *Plant Biology,* 10: 399-408.

[34] Fallahpour, F., Ghorbani, R., Nassiri Mahallati, M. & Hosseini, M. (2015) Demographic Parameters of *Lipaphis erysimi* on Canola Cultivars under Different Nitrogen Fertilization Regimes, Journal of Agricultural Science Technology, 17: 35-47.

[35] Karimi-Malati, A., Fathipour, Y., Talebi, A. A. & Bazoubandi, M. (2014) Life Table Parameters and Survivorship of *Spodoptera exigua* (Lepidoptera: Noctuidae) at Constant Temperatures. Environmental Entomology, 43(**3**): 795-803.

[36] Nikooei, M., Fathipour, Y., Jalali Javaran, M. & Soufbaf, M. (2015) How

Different Genetically Manipulated Brassica Genotypes Affect Life Table Parameters of *Plutella xylostella* (Lepidoptera: Plutellidae). Journal of Economic Entomology, 108(**2**): 515-524.

[37] Taghizadeh, R., Talebi, A. A., Fathipour, Y. & Khalghani, J. (2012) Effect of Ten Soybean Cultivars on Development and Reproduction of Lima Bean Pod Borer, *Etiella zinckenella* (Lepidoptera: Pyralidae) under Laboratory Conditions. Applied Entomology & Phytopathology, 79(2): 15-28

[38] Wittmeyer, J. L. & Coudron, T. A., (2001) Life Table Parameters, Reproductive Rate, Intrinsic Rate of Increase, and Estimated Cost of Rearing *Podisus maculiventris* (Heteroptera: Pentatomidae) on an Artificial Diet. Journal of Economic Entomology, 94: 1344-1352.

[39] Akandeh, M., Kocheili, F., Soufbaf, M., Rasekh, A. & Mozafari, K. (2016) Effect of Canola Physical Mutation on *Plutella xylostella* (L.) Life table, *Journal of Agriculture* Science and Technology, 18: 985-998

[40] Ebrahimi, N., Talebi, A. A., Fathipour, Y. & Zamani, A. A. (2008) Host Plants Effect on Preference, Development and Reproduction of *Plutella xylostella* (L.) (Lepidoptera: Plutellidae) under Laboratory Conditions, *Advances in Environmental* Biology, 2(3): 108-114.

[41] Fathi, S.A.A., Bozorg-Amirkalaee, M. and Sarfaraz, R.M. (2011a) Preference and Performance of *Plutella xylostella* (L.) (Lepidoptera: Plutellidae) on Canola Cultivars. Journal of Pest Science, 84: 41-47.

[42] Fathi, S. A. A. (2010) Host Preference and Life Cycle Parameters of *Chromatomya horticola* Goureau (Diptera: Agromyzidae) on Canola Cultivars. Munis Entomology and Zoology, 5: 247-252.

[43] Fathi, S. A. A., Gholami, F., NouriGanbalani, G. & Mohseni, A. (2011b) Life History Parameters of Thrips tabaci (Thysanoptera: Thripidae) on Six Commercial Cultivars of Canola. Applied Entomology of Zoology, 46: 505-510

[44] Mirmohammadi, S., Allahyari, H., Nematollahi, M. R. & Saboori, A. R. (2009)Effect of Host Plant on Biology and Life Table Parameters of *Brevicoryne brassicae* (Hemiptera: Aphididae). Annals of the Entomology Society of America, 102: 450-455

[45] Ulusoy, M. R. & Olmez-Bayhan, S. (2006) Effect of certain *Brassica* plants on biology of the cabbage aphid *Brevicoryne brassicae* under laboratory. Phytoparasitica 34(2):133-138

[46] Roudposhti, S., Moravej, G. & Hosseini, M. (2012) Evaluation the Resistance of Different Brassica Species to *Lipaphis erysimi* at Greenhouse Conditions, Plant Protection, 26(**2**): 224-230.

[47] Elzinga, D.A. & Jander, G. (2013) The role of protein effectors in plant–aphid interactions, Current opinion in plant biology 16(4): 451-456.

[48] Jaouannet, M., Rodriguez, P.A., Thorpe, P., Lenoir, C.J., MacLeod, R., Escudero-Martinez C. & Bos, J.I. (2014) Plant immunity in plant–aphid interactions, Frontiers in plant science 5: 663

[49] Rodriguez, P.A. & Bos, J.I (2013) Toward understanding the role of aphid effectors in plant infestation, Molecular Plant-Microbe Interactions 26(1): 25-30.

[50] Giordanengo, P., Brunissen, L., Rusterucci, C., Vincent, C., van Bel, A, Dinant, S., Girousse, C., Faucher, M. & Bonnemain, J.L. (2010) Compatible plant–aphid interactions: how aphids manipulate plant responses? Comptes Rendus Biologies 333:516-523.

[51] Firake, D.M., Lytan, D., Behere, G.T., Thakur, N.S.A. (2012a) Host plants alter the reproductive behavior of cabbage butterfly, *Pieris brassicae* (Lepidoptera: Pieridae) and its endo-larval parasitoid, *Hyposoter ebeninus* (Hymenoptera: Ichneumonidae) in cruciferous ecosystems. Florida Entomology, 95(**4**): 905-913.

[52] Firake, D.M., Lytan, D., Behere, G.T. (2012b) Bio-diversity and seasonal activity of arthropod fauna in *brassica* crop ecosystems of Meghalaya, North East India. Molecular Entomology, 3(**4**): 18-22.

[53] Lytan, D. & Firake, D. M. (2012) Effects of different host plants and rearing atmosphere on life cycle of large white cabbage butterfly, *Pieris brassicae* (Linnaeus). Archies of Phytopathology and Plant Protection, 45: 1819-1825.

[54] Thubru, D.P., Firake, D.M. & Behere, G.T. (2016) Assessing risks of pesticides targeting lepidopteran pests in cruciferous ecosystems to eggs parasitoid, *Trichogramma brassicae* (Bezdenko). *Saudi Journal of Biological Sciences*. DOI: http:// dx.doi.org/10.1016/j.sjbs.2016.04.007

[55] Chandra, S. & Kushwaha, K.S.(1981) Management of aphid species infesting mustard, cabbage and cauliflower at Udaipur Ph.D thesis submitted to University of Udaipur.

[56] Dwivedi, S.A. & Singh, R.S. (2020) Synthetic Insecticides and Bio-pesticide Affect Natural Enemies of Aphid (*Lipaphis erysimi* Kalt) in Mustard, *Indian Journal of Agricultural Research* 10.18805/IJARe.A-5613

[57] Rauf, M, Ehsan-ul-Haq Khan, J., Rehman, A., Gillani, W.A. & Ali, A. (2013) Biology and predatory potential of *Coccinella septempunctata* Linn. on *Schizaphis graminum* aphid under controlled conditions. Pakistan Journal of Agricultural Research 26(**2**): 124-129.

[58] Sattar, M., Hamed, M., & Nadeem, S.(2008) Biology of Coccinella septempunctata Linn. (Coleoptera: Coccinellidae) and its predatory potential on cotton aphids, Aphis gossypii Glover (Hemiptera: Aphididae). Pakistan Journal of Zoology, 40(**4**): 239-242.

[59] Srivastava, A.S., Upadhay, K. D., Mishra, B.P. & Katiyar R.R. (1978) Prey preference of *Coccinella septempunctata* L. (Coleoptera :Coccinellidae). Indian journal of Agricultural science **48** (2):84-86.

[60] Manpoong, N.S., Firake, D. M., Behere, G. T. & Rajesh T. (2016) Biological attributes and feeding potential of three dominant predators of *Lipaphis erysimi* (Kaltenbach), Journal of Biological Control, 30(3): 190-194,

[61] Zhang Y.J. & Feng Y.S. (1983) Small scale rearing of *Coccinella Septempunctata* L. and the efficacy of field release. Natural enemies of pest **5**(2):97-99.

[62] Hong, B.M. &Hung H.Q. (2010) Effect of temperature and diet on the life cycle and predatory capacity of *Episyrphus balteatus* (De Geer) (Syrphidae: Diptera) cultured on *Aphis gossypii* (Glover). *Journal of the International Society for Southeast Asian* Agricultural Sciences, 16(**2**): 98-103.

[63] Baskaran, R.K.M., Sasikumar, S., Rajavel, D.S. & Suresh, K. (2009) Biology and predatory potential of aphidophagous syrphid on guava aphid, *Aphis gossypii* (Glover) (Hemiptera: Aphididae). Journal of Biological Control, 23(1): 53-56.

[64] Romabai Devi, Y., Kalita, J., & Singh, T.K. (2011) Biological control potential of an aphidophagous syrphid, *Episyrphus balteatus*, De-Geer (Diptera: Syrphidae) on mustard aphid, *Lipaphis erysimi* (Kalt.) on cabbage ecosystem in Manipur. *Journal of Experimental* Science, 2(**12**): 13-16.

[65] Verma, J.S., Sharma, K.C., Sood A, Sood M. (2005) Biology and predatory potential of syrphid predators on *Aphis fabae* infesting *Solanum nigrum* L. Journal of Entomology Research, 29(**1**): 39-41.

[66] Bakhetia, D.R.C. & Arora, R. (1986) Control of insect pests of toria, sarson and rai. *Indian Farming* 36(4):41-44.

[67] Bakhetia, D.R.C. & Sekhon, B. S. (1989) Insect pests and their management in rapeseed-mustard *Journal of Oilseeds Research* 6:269-273.

[68] Dixon, A.F.G. & Logan, M. (2016) Leaf size and availability of space to the sycamore aphid *Drepanosiphum platanoides*, Acta Oecologica Scandinavica 24: 58-63

[69] Singh, P.K. & Premchand (1995) Yield loss due to mustard aphid, *Lipaphis erysimi* (Kalt.) in Eastern Bihar Plateau. Journal of Applied Zoological Research, 6:97-100.

[70] Srivastava, A. (1999) Effect of date of sowing and varieties on the incidence of mustard aphid, *Lipaphis erysimi* (Kalt.) on rapeseed-mustard. Journal of Oilseeds Research **16** (2): 380-381.

[71] Gupta, R.K. & Agrawal, N (2006) Management of mustard aphid, *Lipaphis erysimi* Kalt. and mustard sawfly, *Athalia proxima* Klug. through different dates of sowing in mustard. Journal of Entomological Research, **30** (4): 325-327.

[72] Kumar, S., Atri, C., Sangha, M.K.& Banga S.S.(2011) Screening of wild crucifers for resistance to mustard aphid, *Lipaphis erysimi* (Kaltenbach) and attempt at introgession of resistance gene (s) from *Brassica fruticulosa* to *Brassica juncea*. Euphytica. 179(3): 461-470.

[73] Choudhury, S., & Pal, S. (2009). Population dynamics of mustard aphid on different Brassica cultivars under terai agro-ecological conditions of West Bengal, Journal of Pharmacy and Pharmaceutical Sciences, *1*(1), 83-86

[74] Dinesh Kumar & Singh, S.P.2012. Screening of *Brassica* germplasm for resistance to mustard aphid, *Lipaphis erysimi* Kalt. based on pooled mean aphid infestation, Annals of Agri Bio Research; 17(2)93-96

[75] Dwivedi, S. A., Singh, R. S. & Pragnabharathi, R (2019) The screening of mustard varieties resistance against mustard aphid *Lipaphis erysimi* Kalt, Plant Cell Biotechnology and Molecular Biology 20(9&10):397-408.

[76] Gupta, M.P., Verma, M.L., Churasia, S.K. (2003) Assessment of avoidable yield losses in Karan Rai varieties due to mustard aphid, Annals of Plant Protection Science, 11: 11-15

[77] Banerjee, A., Datta, J.K. & Mondal, N.K. (2012) Biochemical changes in leaves of mustard under the influence of different fertilizers and cycocel. Journal of Agricultural Technology 8(4): 1397-1411.

[78] Jat, B.L. (2020) Evaluation of yellow sticky trap and imidacloprid 17.8% SL against mustard aphid on farmer's field. *Bhartiya Krishi Anusandhan Patrika,* (35):193-196

[79] Singh, Y. P. & Sharma, K. C. (2009) Monitoring of alate mustard aphid, *Lipaphis erysimi* (Kalt) on yellow sticky traps and its relation to abiotic factors, Indian Journal of Applied Entomology 23(1) : 49-52

[80] Singh, Y. P.& Nagar R. (2016) Monitoring of alate mustard aphid, *Lipaphis erysimi* Kalt on yellow sticky traps, Indian Journal of Ecology 43(1):137-140

[81] Faria, M.R & Wraight, S.P. (2007) Mycoinsecticides and Mycoacaricides: A comprehensive list with worldwide

coverage and international classification of formulation types. Biological Control, 43: 237-256.

[82] Rehner, S.A. (2005) Phylogenetics of the insect pathogenic genus *Beauveria*. pp. 3-27. In: Insect-Fungal Associations: Ecology and Evolution, (Eds.): F.E. Vega and M. Blackwell, Oxford University Press London.

[83] Roberts, D.W. & St. Leger, R.J. (2004) *Metarhizium spp.*, cosmopolitan insect-pathogenic fungi: Mycological aspects. Advances in Applied Microbiology, 54: 1-70.

[84] Asi, M.R., Bahir, M.H., Afzal, M. & Saqib (2009a) Effect of conidial concentration of entomopathogenic fungi on mortality of cabbage aphid, *Brevicoryne brassicae* L. *Pakistan Journal of Life and Social Scienc*e, 2: 175-180

[85] Asi, M.R., Bashir, M.H., Mirza, J. H., Afzal, M. & Imran (2009b) In vitro efficacy of entomopathogenic fungi against cabbage aphid, *Brevicoryne Brassicae* L. Pakistan Entomologist 31(1): 43-47.

[86] Khetan, S.K. (2001) Microbial pest control. Marcel Dekker, Inc. NY, USA..

[87] Meitkiewski, R.T., Pell, J.K. & Clark, S.J. (1997) Influence of pesticide use on the natural occurrence of entomopathogenic fungi in arable soils in the UK: Field and laboratory comparisons. Biocontrol Science Technology, 565-757

[88] Pedro, M.O.J., Neves, E.H., Paulo, T.T. & Alcides, M.J.R. (2001) Compatibility of entomopathogenic fungi with neonicotinoid insecticides, Neotropical Entomology, 30(**2**): 263-268.

[89] Ahmed, S., Ashraf, M.R. & Hussain, A. (2009) Pathogenicity of isolates of *Metarhizium anisopliae* from Gujranwala (Pakistan) against *Coptotermes heimi* (Wasmann) (Isoptera: Rhinotermitidae). *Journal International of Agriculture and Biology*, 11- 6: 707-711.

[90] Liz, J., Mampallil, M.H., Faizal, & Anith, K.N. (2017) Bacterial bioagents for insect pest management. *Journal of Entomology and Zoology Studies*; 5(**6**):2237-2244

[91] Srivastava, A. & Guleria, S. (2003) Evaluation of botanicals for mustard aphid, *Lipaphis erysimi* (Kalt.) control in Brassica, *Himachal* Journal of Agricultural Research *Vol. 29 (1&2): 116-118*

[92] Chitra, K.C., Janardhan R., Kameswara, R. & Nagaiah, K. (1993) Field evaluation of certain plant products in the control of Brinjal pest complex. Indian Journal of Entomology 55 (3): 237-240